PHalarope Books

PHalarope books are designed specifically for the amateur naturalist. These volumes represent excellence in natural history publishing. Most books in the PHalarope series are based on a nature course or program at the college or adult education level or are sponsored by a museum or nature center. Each PHalarope book reflects the author's teaching ability as well as writing ability. Among the books:

Biography of a Planet: Geology, Astronomy, and the Evolution of Life on Earth
Chet Raymo

The Curious Naturalist
John Mitchell and the Massachusetts Audubon Society

A Field Guide to the Familiar: Learning to Observe the Natural World
Gale Lawrence

Botany in the Field: An Introduction to Plant Communities for the Amateur Naturalist
Jane Scott

The Seaside Naturalist: A Guide to Nature Study at the Seashore
Deborah A. Coulombe

Suburban Wildflowers: An Introduction to the Common Wildflowers of Your Back Yard and Local Park
Richard Headstrom

Suburban Wildlife: An Introduction to the Common Animals of Your Back Yard and Local Park
Richard Headstrom

Trees: An Introduction to Trees and Forest Ecology for the Amateur Naturalist
Laurence C. Walker

Thoreau's Method: A Handbook for Nature Study
David Pepi

Wood Notes: A Companion and Guide for Birdwatchers
Richard H. Wood

Insect Life: A Field Entomology Manual for the Amateur Naturalist
Ross H. Arnett, Jr., and Richard L. Jacques, Jr.

Pond and Brook: A Guide to Nature Study in Freshwater Environments
Michael J. Caduto

Illustrations by Charles Douglas

A Natural History Notebook of North American Animals

National Museum of Natural Sciences
National Museums of Canada

 A Spectrum Book

Prentice-Hall, Inc., Englewood Cliffs, New Jersey 07632

Library of Congress Cataloging in Publication Data
Main entry under title:

A natural history notebook of North American animals.

(PHalarope books)
"A Spectrum Book."
Includes index.
1. Zoology—North America.   I. National Museum of Natural Sciences (Canada)   II. Series.
QL715.N38   1985   596.097   85-6307
ISBN 0-13-610114-3
ISBN 0-13-610106-2 (pbk.)

ISBN 0-13-610114-3

ISBN 0-13-610106-2 {PBK.}

© 1985 by Prentice-Hall, Inc., Englewood Cliffs, New Jersey 07632. All rights reserved. No part of this book may be reproduced in any form or by any means without permission in writing from the publisher. A Spectrum Book. Printed in the United States of America.

Illustrations are courtesy of the Natural History Notebook Series, Volumes 1–5, published by the National Museum of Natural Sciences.

10 9 8 7 6 5 4 3 2 1

Editorial/production supervision: Marlys Lehmann
Cover design: Hal Siegel

This book is available at a special discount when ordered in bulk quantities. Contact Prentice-Hall, Inc., General Publishing Division, Special Sales, Englewood Cliffs, N.J. 07632.

Prentice-Hall International (UK) Limited, *London*
Prentice-Hall of Australia Pty. Limited, *Sydney*
Prentice-Hall Canada Inc., *Toronto*
Prentice-Hall Hispanoamericana, S.A., *Mexico*
Prentice-Hall of India Private Limited, *New Delhi*
Prentice-Hall of Japan, Inc., *Tokyo*
Prentice-Hall of Southeast Asia Pte. Ltd., *Singapore*
Whitehall Books Limited, *Wellington, New Zealand*
Editora Prentice-Hall do Brasil Ltda., *Rio de Janeiro*

CONTENTS

FOREWORD xiii

1. An Album of the Past: Prehistoric Animals and Extinct Wildlife of North America 1

LOBE-FINNED FISH 3
COAL-AGE REPTILES 4
HYPACROSAURUS 5
STENONYCHOSAURUS 6
STYRACOSAURUS 7
EUOPLOCEPHALUS 8
TYRANNOSAURUS 9
FLYING REPTILES 10
PANOPLOSAURUS 11
BRACHIOSAURUS 12
DROMICEIOMIMUS 13
THE DAWN HORSE OR EOHIPPUS 14
WESTERN CAMEL 15
YUKON WILD ASS 16
SAIGA ANTELOPE 17
HELMETED MUSKOX 18
ICE AGE ANIMALS ON VANCOUVER ISLAND 19
THE WHITE LAKE WHALE 20
RINGED SEALS 21
AMERICAN MASTODON 22
BABINE LAKE MAMMOTH 23
SCIMITAR CAT 24
WOOLLY MAMMOTH 25
SHORT-FACED BEAR 26
JEFFERSON'S GROUND SLOTH 27
AMERICAN LION 28
EARLIEST DOGS IN NORTH AMERICA 29
PASSENGER PIGEON 30
CAROLINA PARAKEET 31
GREAT AUK 32
GIANT BEAVER 33

2. Animals of North America Today 35

Section One
FISHES/AMPHIBIANS/REPTILES 37

SHARKS 38
WHALE SHARK 39
BLUE SHARK 40
CARP 41
NORTHERN PIKE 42
SEA LAMPREY 43
BROWN BULLHEAD 44
SHORTNOSE STURGEON 45
ATLANTIC SALMON 46
BROWN TROUT 47
YELLOW-SPOTTED SALAMANDER 48
NORTHERN LEOPARD FROG 49
STRIPED CHORUS FROG 50
BULLFROG 51
RACER 52
MASSASAUGA RATTLESNAKE 53
COMMON GARTER SNAKE 54
BLACK RAT SNAKE 55
SPINY SOFTSHELL TURTLE 56
COMMON SNAPPING TURTLE 57
ATLANTIC RIDLEY 58
ATLANTIC LOGGERHEAD TURTLE 59

LEATHERBACK TURTLE   60
CROCODILE   61
AMERICAN ALLIGATOR   62
ALLIGATOR LIZARD   63

Section Two
BIRDS   65

RAZORBILL/COMMON MURRE   66
NORTHERN GANNET   67
DOVEKIE   68
ATLANTIC PUFFIN   69
GREATER PRAIRIE-CHICKEN   70
AMERICAN WHITE PELICAN   71
CANADA GOOSE   72
TRUMPETER SWAN   73
WHOOPING CRANE   74
GREAT BLUE HERON   75
CONDOR   76
BURROWING OWL   77
SNOWY OWL   78
PEREGRINE FALCON   79
GOLDEN EAGLE   80
BALD EAGLE   81
PIPING PLOVER   82
YELLOW-BELLIED SAPSUCKER   83
COMMON NIGHTHAWK   84
NORTHERN SHRIKE   85
SNOW BUNTING   86
BARN SWALLOW   87
RUBY-THROATED HUMMINGBIRD   88
AMERICAN CROW   89
COMMON RAVEN   90
ESKIMO CURLEW   91
AMERICAN ROBIN   92
KIRTLAND'S WARBLER   93

Section Three
MAMMALS   95

EASTERN CHIPMUNK   96
NORTHERN FLYING SQUIRREL   97
RICHARDSON'S GROUND SQUIRREL   98
RED SQUIRREL   99
HOUSE MOUSE   100
KANGAROO RAT   101
BROWN RAT   102
VANCOUVER ISLAND MARMOT   103
MUSKRAT   104
BEAVER   105
AMERICAN PIKA   106
HOARY MARMOT   107
ARCTIC HARE   108
AMERICAN PORCUPINE   109
STAR-NOSED MOLE   110
HAIRY-TAILED MOLE   111
OTTER   112
SEA OTTER   113
AMERICAN MINK   114
LEAST WEASEL   115
ERMINE   116
BLACK-FOOTED FERRET   117
AMERICAN MARTEN   118
AMERICAN BADGER   119
WOLVERINE   120
STRIPED SKUNK   121
RACCOON   122
RED FOX   123
COYOTE   124
WOLF   125
ARCTIC WOLF   126
LYNX   127
BOBCAT   128

OCELOT   129
JAGUAR   130
COUGAR   131
BLACK BEAR   132
GRIZZLY BEAR   133
POLAR BEAR   134
BIG BROWN BAT   135
PRONGHORN ANTELOPE   136
MUSKOX   137
MOUNTAIN GOAT   138
BIGHORN SHEEP   139
ELK   140
PEARY CARIBOU   141
MOOSE   142
BISON   143
WALRUS   144
NORTHERN ELEPHANT SEAL   145
HOODED SEAL   146
HARP SEAL   147
GRAY WHALE   148
BOWHEAD WHALE   149
FIN WHALE   150
HUMPBACK WHALE   151
SEI WHALE   152
BLUE WHALE   153
SPERM WHALE   154
WHITE WHALE   155
ORCA, OR KILLER WHALE   156
FLORIDA MANATEE   157

BIBLIOGRAPHY—SELECTED READING
   LIST   159
INDEX   161

Foreword

Who has not wondered about the animals that today are found only as fossils in a museum or buried in the ground and asked what they were really like? How did they walk or crawl? What did they eat? What creatures were their enemies? What about the animals that are alive today? Often we hear their names or are told in guidebooks what the exact features are to identify them, but it is difficult to find something about their life styles.

A Natural History Notebook of North American Animals is based on a series originally published by this museum. It presents a capsule of information, in easy-to-understand language and pictures, about animals that live now or that once lived in North America. In addition to the natural history of each animal presented, for those animals that are living today, there are often hints on how to observe them more easily in their natural habitat. Many of the animals described are "endangered species," fighting for their survival.

If you are interested in this book, or its predecessor, the Natural History Notebook series, you are probably among the amateur naturalists in North America who feel we should protect our natural environment and the creatures who live in it. Canada's National Museum of Natural Sciences takes its role as educator very seriously. We feel that the more people who know and understand about our wildlife and its needs, the better our own chances for survival in this natural world will be.

The Natural History Notebook series and the present volume were prepared by educators and curators (museum scientists) at Canada's National Museum of Natural Sciences. Originally, descriptions of each animal were printed separately in nearly 200 newspapers, as well as weekly and monthly reviews in Canada, the United States, and Africa. The form it takes in this book now brings much of the information together in one place for easy reference. It provides just a brief sampling of the wealth and variety of American and Canadian wildlife.

I hope A Natural History Notebook of North American Animals will be interesting and informative to you, as the series has been to many others, and that it will help increase your knowledge and appreciation of the wonderful world around us.

Alan R. Emery, director
National Museum of Natural Sciences
Ottawa, Canada

1

An Album of the Past: Prehistoric Animals and Extinct Wildlife of North America

LOBE-FINNED FISH

By 360 million years ago, movements in the earth's crust had brought Scandinavia and the Maritime provinces of Canada together, forming a great north-south chain of mountains between them. Seasonal rains swept mud and sand into a broad, shallow lake in eastern Quebec, which then lay on the equator. These deposits, which can be seen at the Musée d'Histoire Naturelle de Miguasha, contain the most diversified and excellently preserved fish fossils from the period, including this famous lobe-fin Eusthenopteron.

Juveniles inhabited quiet, deeper waters while adults, with their more powerful tail fins, frequented the shores. Related to the ancestors of the amphibians, they used their muscular paired fins to move through the shallows and to raise their heads above water to breathe.

COAL-AGE REPTILES

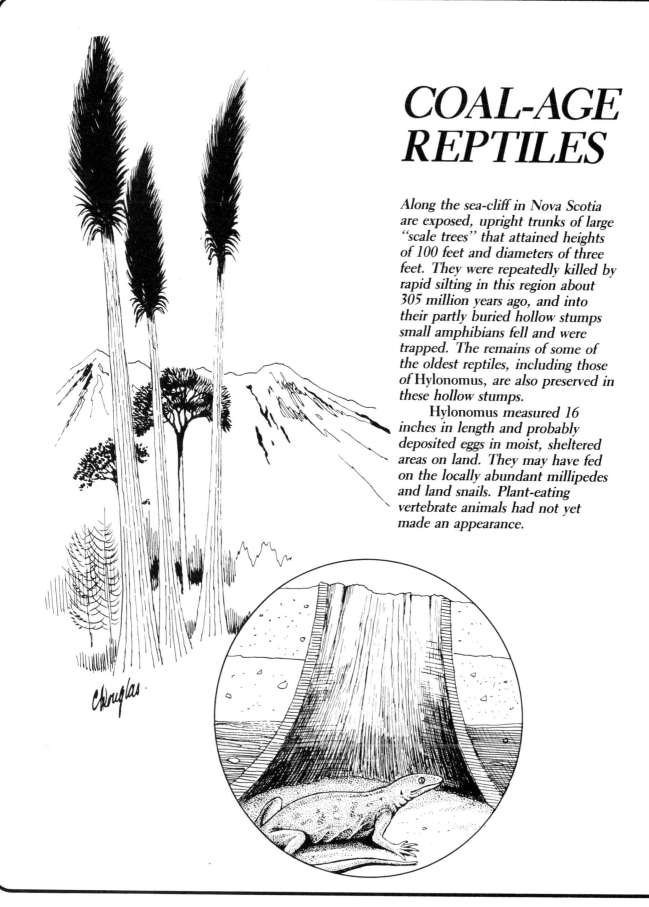

Along the sea-cliff in Nova Scotia are exposed, upright trunks of large "scale trees" that attained heights of 100 feet and diameters of three feet. They were repeatedly killed by rapid silting in this region about 305 million years ago, and into their partly buried hollow stumps small amphibians fell and were trapped. The remains of some of the oldest reptiles, including those of Hylonomus, are also preserved in these hollow stumps.

Hylonomus measured 16 inches in length and probably deposited eggs in moist, sheltered areas on land. They may have fed on the locally abundant millipedes and land snails. Plant-eating vertebrate animals had not yet made an appearance.

HYPACROSAURUS.

Duck-billed dinosaurs were the most numerous plant-eating dinosaurs in the northern hemisphere during the latter part of the age of reptiles. Skeletal material has also been found in Argentina. Duck-billed dinosaurs had grinding, elephantlike dentitions and seem to have appeared at the same time as woody, flowering plants.

Their presence in the Arctic suggests summer migrations into then-warm polar regions with the arrival of summer daylight. Many different kinds of duck-billed dinosaurs are known. Hypacrosaurus possessed a hollow hood on its head, which has been interpreted as a water-trap, a smelling organ, a resonating chamber for trumpeting, a visual display organ, a humidifying device, and an organ to cool the brain.

STENONYCHOSAURUS

Among the rarest of dinosaurian fossils are those of the smaller flesh eaters. One of these, from 76-million-year-old sediments in Dinosaur Provincial Park, Alberta, Canada, is unusually interesting. Stenonychosaurus was less than ten feet long, and weighed only 100 pounds. It was a biped and could rotate the lower arm to grasp objects with a three-fingered hand. Stenonychosaurus' eyes were enormous, surpassing in size those of most modern land animals. It probably fed on primitive mammals.

    This dinosaur's brain was much larger than those of today's reptiles and in relative size approached that of some of today's birds and mammals. Stenonychosaurus, which probably fed on primitive mammals, embodies a widespread tendency for the brain to increase in size through the history of life.

STYRACOSAURUS

Styracosaurus, *a familiar horned dinosaur of moderate proportions, is known only from the 76-million-year-old strata of Dinosaur Provincial Park, Alberta, Canada.*

There, a few articulated specimens have been discovered, as well as concentrations of tumbled Styracosaurus bones, which are suggestive of mass death. Resembling the rhinoceros in general appearance, horned dinosaurs did not hold their forelimbs directly beneath their bodies, and their jaws were adapted to chop, rather than grind, plant food.

Many different horned dinosaurs of varying sizes and horn shapes are known to have lived in ancient North America. Their ability to defend themselves is reflected in the presence of a bony lattice-work protecting the stomachs of the tyrannosaurs.

EUOPLOCEPHALUS

Remains of armored dinosaurs that possessed tail clubs were first discovered in North America, although a greater variety is now known from central Asia. Euoplocephalus, a ponderous, clumsy creature weighing 2.75 tons, occurred in some abundance in Alberta, Canada. It was a squat animal whose trunk was bound by a nearly immovable cuirass of bony armor. Even the eyelids were protected by movable, bony plates.

The suppleness of its forelimbs suggests Euoplocephalus could dig, and complicated cranial sinuses suggest a keen sense of smell. The small brain and weak jaws complement the general impression of a sedentary animal that fed on fleshy vegetation and tubers close to the surface of the ground.

TYRANNOSAURUS

Remains of a large carnivorous dinosaur were named Tyrannosaurus in 1905, and the name has since become a household word. Only about a dozen specimens consisting of more than one bone each are known, but isolated fragments occur in 65-million-year-old sediments from Texas to Canada. The most complete specimen is a skeleton lacking the limbs, which, combined with legs from another individual, is on display in the American Museum of Natural History in New York.

Tyrannosaurus, the largest known tyrannosaur, weighed about 4.4 tons. Smaller tyrannosaurs have also been found in North America and Asia.

These animals were active, aggressive carnivores that either bolted large pieces of their prey, or swallowed them whole.

FLYING REPTILES

Although their fossil remains are not often found, pterosaurs, or flying reptiles, were probably the most abundant winged, vertebrate animals of the age of dinosaurs. The jaws of Pterodaustro, one of the most peculiar pterosaurs known, were lined with hundreds of long, slender teeth. These teeth were apparently used to filter microorganisms from the water. Pterodaustro remains have been excavated in some abundance from 70-million-year-old freshwater deposits in central Argentina.

With a wingspan of eight feet, the animal was much smaller than the largest pterosaurs, which possessed wingspans of up to 51 feet. The structure of giant pterosaurs suggests they were adapted to flying in an atmosphere denser than our present atmosphere.

PANOPLOSAURUS

Two related groups of large, armored dinosaurs are known to have existed during the latter part of the age of reptiles. They were of similar body size and proportions, and in both groups the upper part of the body was covered with saucer-like, bony plates embedded in the skin. However, in the group to which Panoplosaurus belongs, there was no bony club-like mass at the end of the tail; instead, the animal was protected by a row of great, solid spines located on its flanks.

Panoplosaurus remains are not abundant, although three species are known from western Canada. Smaller, related forms occur in Europe, but these particular dinosaurs are unknown elsewhere.

BRACHIOSAURUS

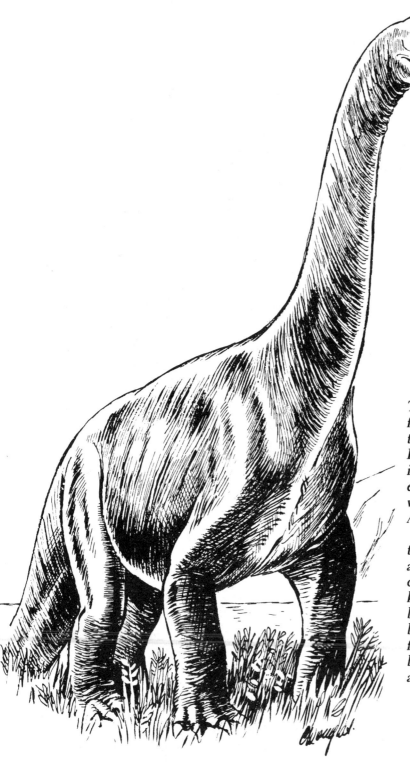

This huge herbivore, which was 50 feet long and weighed 50 tons, was the largest known land animal. Fossil evidence suggests that it inhabited tropical forest riverbank environments in what are now the west-central United States, East Africa, and Portugal.

The era of the dinosaurs came to an end about 65 million years ago, but the Brachiosaurus had died out before this. It is not known why dinosaurs disappeared, but at the time there was a global biological stress and many other forms of animals and plants also became extinct. Man did not appear until millions of years later.

DROMICEIOMIMUS

Occurrences of the lightly constructed skeletal parts of ostrich dinosaurs are tantalizingly incomplete, but have been recovered from Africa, Asia, Europe, and North America. They span the final 75 million years of the age of reptiles. As in the case of the horse, the history of ostrich dinosaurs demonstrates an increasingly excellent adaptation to running. Some species would clearly have rivaled the modern ostrich in speed. Dromiceiomimus, an advanced form, lived in western Canada between 75 and 80 million years ago; the Dromiceiomimus' brain was comparable in size to that of an ostrich. But, like most ostrich dinosaurs, Dromiceiomimus possessed long forelimbs, which were used to uncover small animals and eggs near the surface of the ground.

THE DAWN HORSE OR EOHIPPUS
HYRACOTHERIUM

These small ancestors of modern horses were 19 inches or less in length—about the size of a fox terrier. When compared with modern horses, their legs were shorter, their heads were longer in relation to their bodies, and they had a more complete series of teeth. They had three toes on their hind feet and four on their forefeet. Each toe had a pad on its underside, as do dogs' paws. Modern horses have long legs, each of which ends in a single, powerful toe with a hoof, but no pad. Eohippus lived during the early part of the Tertiary Period (about 50 million years ago). Although these dawn horses were present in Europe as well as North America, the major part of horse evolution occurred on the latter continent.

WESTERN CAMEL
CAMELOPS HESTERNUS

These camelids evidently looked like living dromedaries (single-humped camels), but they were about one fifth taller. During the ice age, herds ranged from Alaska and the Yukon throughout western Canada and the United States to Mexico.

They were the most widely distributed camelids in North America.

Some specimens are very well preserved: One from a cave in Utah still had ligaments attached after 11,000 years of burial.

Paleo-Indians occasionally hunted western camels in Wyoming 10,000 years ago, which is approximately when the species died out.

YUKON WILD ASS
EQUUS (ASINUS) LAMBEI

These small horses were one of the most common of the larger animals in northwestern North America during the ice age. They were probably very much like the living wild asses of Asia, and stood over four feet high. In Canada, most specimens (including some well-preserved skulls) are from the Dawson area of the Yukon. Other fossils have been found in the Northwest Territories and Alaska.

The species probably occupied rather dry, shrub-covered grasslands, as suggested by both the range of living wild asses and the frequency of rootlet impressions on the surfaces of many of the fossil bones. Probably only the fastest and craftiest predators, such as the American lion and wolf, were able to kill them. Wild asses survived in the Yukon until about 13,000 years ago.

SAIGA ANTELOPE
SAIGA TATARICA

Only one fossil of this species is known from Canada. It was found on the beach of Baillie Islands, east of the MacKenzie delta in the Northwest Territories. Other remains have been found in ice-age deposits of central and northern Alaska. Because saiga antelopes are particularly adapted to dry, steppe-grasslands, they probably crossed the wide steppe-like plains of the Bering Isthmus (a land connection between Siberia and Alaska that existed during glacial phases of the ice age) during the last glaciation. These small, slender-limbed mammals with characteristic swollen muzzles became extinct in North America about 10,000 years ago, but saiga herds still survive in the dry grasslands of central Asia.

HELMETED MUSKOX
SYMBOS CAVIFRONS

According to the number of fossils recovered, this was the most common muskox in North America during the last glaciation. The Symbos was taller and had a more slender build than the modern tundra muskox. Its longer, deeper skull supported higher, more flaring horns with massive, fused bases. Symbos fossils have been found from Alaska and the Yukon to New Mexico, and from Vancouver Island to the Atlantic coast. Hair and pellet-like winter droppings found with a well-preserved skeleton from Alaska indicate that helmeted muskoxen had dark brown coats and fed on grasses and sedges. The species died out about 11,000 years ago.

ICE-AGE ANIMALS ON VANCOUVER ISLAND

Herds of imperial and (possibly) Columbian mammoths, American mastodons, helmeted muskoxen, horses, and bison once lived on southeastern Vancouver Island. Most of the approximately 20 fossils representing these species are from gravel pits in the Saanich Peninsula. Others are from the Shawinigan Lake and Courtenay areas.

Most of these land mammals probably reached the island in front of southerly advancing glacial ice some 20,000 to 30,000 years ago by crossing large flood plains that filled the Strait of Georgia. The Strait, like the Bering Sea, was solid land at the time, enabling animals to cross easily. However, the Vancouver Island marmot (Marmota vancouverensis), which is adapted to alpine meadows, may be a living representative of an earlier crossing. Presumably it (the only species of mammal unique to the island) arose from hoary marmots of the mainland that reached what is now Vancouver Island during the next-to-last glaciation.

THE WHITE LAKE WHALE
BALAENA MYSTICETUS

As the last ice sheet retreated north of the St. Lawrence River valley some 12,000 years ago, the Atlantic Ocean flooded the depressed lowland and created the Champlain Sea. Bones of a large bowhead whale (which probably averaged about 52 feet in length and weighed 55 tons in life) were found in beach deposits of the Champlain Sea near the town of White Lake, west of Ottawa.

A radiocarbon dating of bone from the specimen suggests that this arctic-adapted whale died and was washed ashore, or was stranded and died, about 11,500 years ago. Other remains of large whales that occupied the Champlain Sea have been found near Smith Falls, Ontario, and Daveluyville and Les Cèdres, in Quebec.

RINGED SEALS
PHOCA (PUSA) HISPIDA

These arctic-adapted seals lived in the western part of the Champlain Sea about 11,000 years ago. As the last ice sheet retreated north of the St. Lawrence River valley some 12,000 years ago, the Atlantic Ocean flooded the depressed lowland, creating this inland sea.

Most of a skeleton was collected from Champlain Sea clay near Hull, Quebec, together with shells of a small clam (Hiatella arctica). The shells indicate the deposition in cool (32° to 50°F), salty water. As ringed seals are adapted to keeping breathing holes open in sea ice in winter and to pupping on land-fast ice in spring, those conditions presumably existed in the western part of the Champlain Sea during its early stages. The ringed seal skeleton is on display at the National Museum of Natural Sciences in Ottawa, Canada.

AMERICAN MASTODON
MAMMUT AMERICANUM

When compared with modern elephants, mastodons were squat and long in the body. Vestigial tusks were often present in their lower jaws. Shoulder height varied from over six feet to almost nine feet. Their hair was coarse and reddish brown. This species was confined to North America, being most common in the east where it found favorable habitat in spruce forest or open spruce woodland. Mastodons are known to have eaten conifer twigs and cones, leaves, grasses, and swamp plants. Scimitar cats evidently hunted young mastodons, while adults were sometimes killed by Paleo-Indians. Mastodons lived throughout the ice age (the last two million years), becoming extinct about 4,000 years ago.

BABINE LAKE MAMMOTH
MAMMUTHUS CF. COLUMBI

In 1971, workmen excavating in an open-pit copper mine at Babine Lake, in central British Columbia, discovered the partly articulated skeleton of a Columbian mammoth.

    The bones were taken from silty pond deposits overlain by very thick boulder-clay, which was deposited by the last glacier that covered the area. Radiocarbon dating indicates that the animal sank in sticky pond deposits about 34,000 years ago, and died there.

    A study of plant remains found with the bones indicates that the vegetation near Babine Lake about the time the mammoth lived was similar to shrub tundra now found beyond the tree line in northern Canada.

SCIMITAR CAT
HOMOTHERIUM SERUM

Scimitar cats, related to the better known saber-toothed tigers, were about the size of a lion and had razor-sharp stabbing teeth. They had a bob tail like the lynx, long and powerful forelimbs, and poorly developed hind limbs.

Scimitar cats seem to have specialized in killing young mammoths and probably other slow, thick-skinned prey, such as mastodons and ground sloths. They were once widespread in North America. In Canada, their remains are known from ice-age deposits in the western Yukon. Friesenhahn Cave, Texas, is probably the most abundant source of scimitar cat remains in North America. Scimitar cats survived in North America and England until the close of the last glaciation, about 10,000 years ago.

WOOLLY MAMMOTH
MAMMUTHUS PRIMIGENIUS

Woolly mammoths are perhaps the best-known mammals of the ice age. Much is known about their appearance as their carcasses have been found preserved in frozen ground in Siberia, and wall pictures by stone-age artists can be seen today in ancient European caves.

Woolly mammoths grew to about the size of present-day Asiatic elephants. They possessed warm coats consisting of long, brown guard hairs and soft underwool. They had large, curved ivory tusks and knob-like heads.

Woolly mammoths once roamed the northern parts of Eurasia and North America, feeding on plants such as grasses, sedges, and shrubs. One of the best-preserved North American specimens consists of almost an entire skeleton from the Whitestone River in the Yukon Territory. Radiocarbon dating indicates that the animal died there about 30,000 years ago. Their thick coats and heavy layers of fat fitted them for life in cold, tundra-like conditions. Primitive hunters were sometimes successful in killing them for food.

SHORT-FACED BEAR
ARCTODUS SIMUS

This bear seems to have been mainly a flesh-eater, and was by far the most powerful land predator during the ice age in North America. It may have attacked bison, deer, and horses. The largest known skull of Arctodus was found by a Yukon gold miner. Another fossil from southern Saskatchewan indicates that Arctodus lived there more than 70,000 years ago. This species ranged the high grasslands of western North America from Alaska to Mexico, while a more lightly built species (Arctodus pristinus), with smaller teeth, preferred the more heavily wooded Atlantic coastal region. The short-faced bear became extinct some 10,000 years ago, perhaps partly because some of its large prey died out earlier, and possibly because of competition with the smaller, more herbivorous brown bears (Ursus arctos) that entered North America from Eurasia.

JEFFERSON'S GROUND SLOTH
MEGALONYX JEFFERSONII

Ground sloths were distant relatives of tree sloths, which now live in the forests of South America. They are among the most unusual of North American ice-age mammals. The long-haired, bear-sized Megalonyx may have reached northwestern North America from the south during a relatively warm period more than 70,000 years ago. In North America, fossils have been found in the Yukon, the Northwest Territories, western Canada, and many of the states. The ground sloth's teeth suggest it fed on tree foliage and shrubs. This species evidently became extinct about 9,000 years ago.

AMERICAN LION
PANTHERA LEO ATROX

Lions, slightly larger than living African lions, once roamed the Americas from Alaska to Peru.

Their remains, ranging in age from 20,000 to more than 70,000 years, have been found in ice-age deposits in both the western Yukon and southern Alberta, Canada.

These predators were perhaps best adapted to hunting large-horned bison and wild horses in open grasslands and parklands. They became extinct about 10,000 years ago, following the decline of their large prey. Paleolithic art from France and the Soviet Union suggests that the closely related "cave lions" of Eurasia had faintly striped coats, so American lions may have had a similar appearance.

EARLIEST DOGS IN NORTH AMERICA
CANIS FAMILIARIS

Dog remains are often found in Indian and Inuit archaeological sites.

Indeed, Thule people, ancestors of the Inuit, used sled dogs in the North American Arctic some 1,000 years ago. However, there are much earlier records indicating that dogs of two distinct sizes lived near Jaguar Cave, Idaho, about 10,000 years ago.

Different breeds may have existed toward the close of the last glaciation, but the earliest known dogs seem to be from the Yukon.

A deeply stained jaw from the Old Crow basin may indicate that small, long-snouted dogs lived there, with people, more than 20,000 years ago.

PASSENGER PIGEON
ECTOPISTES MIGRATORIUS

Once a common bird of eastern North America, the Passenger Pigeon became rare in the nineteenth century. Early records, beginning in 1630, describe its migrations, roostings, and nestings in enormous numbers, but by 1912 rewards were being offered for evidence of a live, wild bird.

There is a question of the natural survival ability of a bird that roosted in such large numbers that it destroyed forests, that laid one or two eggs in a flimsy nest, and suffered losses from overcrowding and nestling mortality. But man finally doomed the bird to extinction. Shot, trapped, and clubbed for market, hog food, and sport, it could not survive. The last Passenger Pigeon died in a zoo in 1914.

CAROLINA PARAKEET
CONUROPSIS CAROLINENSIS

With the spread of agriculture, this brilliantly colored bird developed a liking for the seeds of many kinds of fruit and grain crops. For this, and its habit of gathering in great, destructive flocks, it was condemned as a pest and subjected to wholesale slaughter. Many were also sold as caged pets.

Although once common in the southeastern United States, it became increasingly scarce as deforestation reduced its habitat. Already rare by the mid-1880s, its last stand was in Florida. There, in 1920, a flock of 30 birds was the last ever seen of this only native parrot of the United States.

GREAT AUK
PINGUINUS IMPENNIS

Large breeding colonies of this flightless, penguinlike sea bird once gathered on rocky islands and coasts of the north Atlantic from Canada to Greenland, Iceland, the British Isles, and Scandinavia.

A strong swimmer, the Great Auk wintered as far south as Florida and southern Spain.

Its extermination began with a slaughter for food and bait by local inhabitants and fishermen, and continued for the bird's fat and feathers.

As the birds became scarce, they were collected for a lucrative trade in skins and eggs. The last known living pair and one egg were taken in Iceland in 1844, and the Great Auk is now represented in collections only by bones, skins, and eggs.

COMPARISON OF PRESENT-DAY BEAVER WITH CASTOROIDES

GIANT BEAVER
CASTOROIDES

The giant beaver (Castoroides ohioensis) *was one of the largest rodents ever known, reaching a length of about eight feet and weighing up to an estimated 480 pounds. Unlike modern beavers* (Castor canadensis), *giant beavers had ridged cutting teeth, deep skulls, and probably roundish, muskratlike tails.*

In Canada, fossils over 70,000 years old have been found in Toronto and in the Old Crow basin of the Yukon.

This animal became extinct, along with mammoths, mastodons, and ice-age horses, about 10,000 years ago.

2
Animals of North America Today

Section One
FISHES / AMPHIBIANS / REPTILES

SHARKS

PACIFIC MAKO SHARK
(Isurus glaucus)
This is one of the most active sharks, leaping repeatedly when hooked, which makes it a popular quarry for shark fishermen. The mako is involved in attacks on boats more frequently than any other species.

GREAT WHITE SHARK
(Carcharodon carcharias) *This is the most dangerous of all the sharks. One of the largest reported to date measured 37 feet and probably weighed fourteen tons. It is very aggressive.*

THRESHER SHARK
(Alopias vulpinus)
Distinguished by the long (up to ten feet) upper lobe of the caudal fin, which is used to strike whiplike blows at its prey.

SMOOTH HAMMERHEAD
(Sphyraena zygaena)
This is one of several hammerhead species. The eyes are located at the outer tips of the head. It grows to lengths of 15 feet or more.

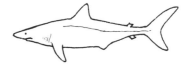

BLUE POINTER OR PACIFIC MAKO
(Isurus glaucus)
Perhaps the most beautiful of the sharks. The back is a deep blue, the underside a glittering white. It grows to weights of 1,000 pounds and up to 12 feet in length.

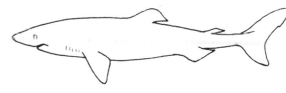

GREENLAND SHARK
(Sōmniosus microcephalus) *A very lethargic creature, and one of the few sharks to inhabit polar waters year-round.*

WHALE SHARK
RHINCODON TYPUS

The whale shark is the largest known fish. It is known to reach 50 feet in length and possibly 60 feet or more. Its weight can exceed ten tons. This creature ranges all tropical waters, and infrequently strays into temperate ones. It is mainly solitary in nature and, despite its impressive appearance, it is harmless to man. Scuba divers and underwater swimmers have clambered unmolested over its body.

The whale shark feeds chiefly on plankton, but it also consumes sardines and anchovies. It is an egg-layer, but to date only one egg case of this great fish has been found. It contained a perfect 14-and-a-half-inch-long replica of its enormous parent.

BLUE SHARK
PRIONACE GLAUCA

The blue shark is one of the most abundant and far-ranging of all sharks, and is a prolific breeder. Litters sometimes contain as many as 70 pups. This is a slimly built fish; a seven-foot specimen weighed only about 70 pounds. Regarded by anglers as a sport fish, the current world angling record of the blue shark is 11 feet, six inches and 410 pounds.

Sometimes known as the blue whaler because of its frequent presence at the scenes of whale kills, the blue shark does not have a reputation, as do some other sharks, of being particularly dangerous to man. However, as with others of its kind, it is unpredictable and has possibly claimed some human victims.

CARP
CYPRINUS CARPIO

The carp was introduced to North America from Europe. Sometimes specimens have only a few large scales ("mirror carp") or none at all ("leather carp").

The carp is known to frequent the warm, shallow waters of lakes and streams, even when they are somewhat muddy or polluted. It feeds on insect larvae, crustaceans, snails, and plants. It spawns in vegetated shallows in June and July, and a 17-pound specimen can spawn 2.3 million eggs. The current world record for carp is a South African specimen that weighed 83.5 pounds when caught. This species is usually caught by still fishing using doughballs, potatoes or worms for bait.

NORTHERN PIKE
ESOX LUCIUS

One of North America's most voracious freshwater predators, pike often consume prey half their own length, including minnows, frogs, crayfish, mice, muskrats, and ducklings, as well as their own kind.

Arctic pike may live for as long as 25 years.

Primarily a freshwater fish, the pike is found in most of Canada except the Maritime provinces, and throughout the north-central United States. The North American angling record is a specimen 46 pounds, two ounces in weight and 51.5 inches in length, although an unconfirmed report exists of a 55-pound pike taken from a lake in Alberta, Canada.

Unlike the muskellunge, hooked pike fight in the depths rather than leaping from the water. As a sport fish, it is usually taken by trolling with large spoons, plugs, large bait fishes, or worm harness.

SEA LAMPREY
PETROMYZON MARINUS

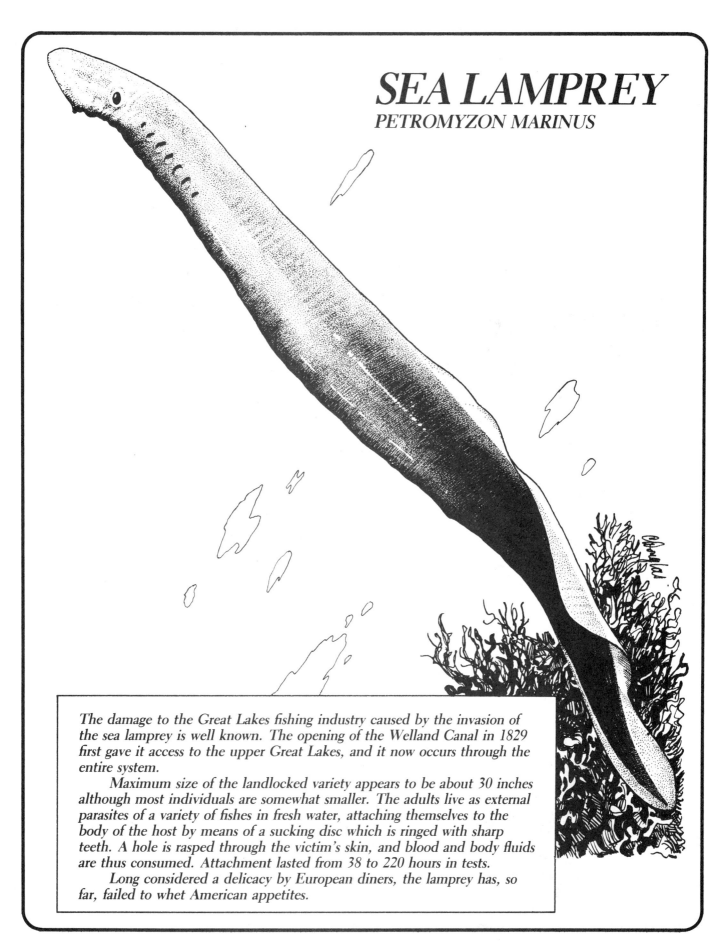

The damage to the Great Lakes fishing industry caused by the invasion of the sea lamprey is well known. The opening of the Welland Canal in 1829 first gave it access to the upper Great Lakes, and it now occurs through the entire system.

Maximum size of the landlocked variety appears to be about 30 inches although most individuals are somewhat smaller. The adults live as external parasites of a variety of fishes in fresh water, attaching themselves to the body of the host by means of a sucking disc which is ringed with sharp teeth. A hole is rasped through the victim's skin, and blood and body fluids are thus consumed. Attachment lasted from 38 to 220 hours in tests.

Long considered a delicacy by European diners, the lamprey has, so far, failed to whet American appetites.

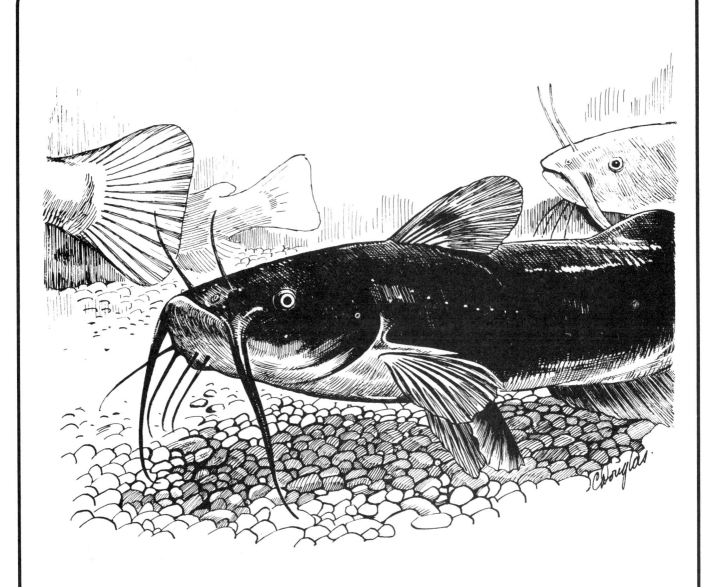

BROWN BULLHEAD
ICTALURUS NEBULOSUS

The normal length of this moderate-sized bullhead is eight to 14 inches. Native distribution is restricted to the fresh waters of eastern and central North America.

This species was released in Germany in the early 1900s and from there spread to England, many European countries, and the USSR. Its maximum age is eight years.

Nests are usually shallow depressions in a muddy or sandy bottom, in which the eggs are deposited. Feeding is done mainly at night, on or near the bottom, and food, including waste and offal, is searched out largely by means of the barbels. The brown bullhead is fairly resistant to domestic and industrial waste, and was the only species found in some heavily polluted streams near Montreal, Quebec.

The flesh is firm, reddish to pink in color, and quite delicious.

SHORTNOSE STURGEON
ACIPENSER BREVIROSTRUM

In the United States, where this fish is classified as endangered, its range consists of tidal rivers and coastal waters from Maine to Florida; however, largely because of pollution, it has disappeared from much of this range. The only Canadian occurrence of this small sturgeon, which can reach up to four feet, nine inches in length, is in the St. John River, New Brunswick.

Although a rare species in Canada, some are caught accidentally in commercial fishing operations, and, according to size limitations, may be marketed.

Recommendations for protection include pollution controls, larger gill net mesh size, and the use of traps instead of gill nets.

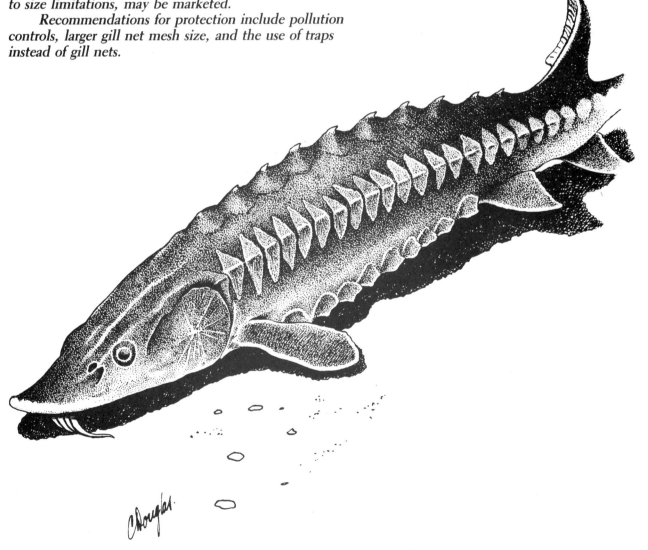

ATLANTIC SALMON
SALMO SALAR

Dams and pollution are hazards for the Atlantic salmon on its runs to the spawning beds, but an earlier and far more serious problem has been high-seas fishing, which was not subject to management regulation for sustaining yields.

In 1966, when reduced stocks caused concern on both sides of the Atlantic, Canada, the United States, and Spain joined in a ban on high-seas salmon fishing, although other countries did not join the ban until ten years later.

Canada has banned all commercial Atlantic salmon fishing since 1972, and has compensated fishermen for their losses. Sport fishing for salmon in Atlantic Canada and Quebec has continued under strict controls.

BROWN TROUT
SALMO TRUTTA

The brown trout was introduced to North America from Europe in 1883 and has since become a popular quarry for anglers.

The females spawn in autumn and the beginning of winter at a water temperature of about 45°F, laying their eggs in shallow water, on gravelly bottoms. A nest is dug in the gravel, where the eggs are deposited and then covered over. A five- or six-year-old female would produce about 2,000 eggs per season.

The world record for this fish is 39 pounds and 3.38 feet in length, taken in Loch Awe, Scotland, in 1866. The best time for fishing for the brown trout is in the evening. It feeds upon aquatic and terrestrial insects, crustaceans, molluscs, frogs, salamanders, and other fishes.

YELLOW-SPOTTED SALAMANDER
AMBYSTOMA MACULATUM

The word "salamander" originates from an ancient belief that these animals were fire-lovers. Perhaps this started when someone saw such a creature, when the log in which it had taken cover was placed in a bonfire, escaping, rather than emerging, from the fire—a significant difference!

Spotted salamanders are rarely seen by most naturalists because these animals spend most of the year beneath forest litter or under the ground, rarely emerging in the daytime.

In early spring, however, soon after the snow melts, they migrate to ponds and ditches, particularly on rainy nights, and breed.

Salamanders are readily recognizable by their black color and their patterns of bright yellow or orange spots.

NORTHERN LEOPARD FROG
RANA PIPIENS

The Northern Leopard Frog ranges from Labrador to the Northwest Territories, south to Pennsylvania and New Mexico, and west to the Pacific states. It is primarily a frog of meadows and other grassy areas in summer.

In early spring it breeds in ponds, marshes, and at the quiet edges of rivers, lakes, and streams. Tadpoles of these frogs generally change in July and August into froglets slightly less than one inch long. They breed first at a length of about two inches but can grow to body lengths of up to 4.25 inches.

STRIPED CHORUS FROG
PSEUDACRIS TRISERIATA

These are small frogs with big voices. The males have a rasping breeding call that sounds like a thumbnail repeatedly drawn over the teeth of a pocket comb, and this is often assumed to come from a much larger frog.

A pattern of longitudinal stripes, spots, or both marks the back and sides. Those found in southern Quebec and southern Ontario southwest to Oklahoma are brown in color, but those found from northeastern Ontario through the Canadian prairies to New Mexico are frequently red or green. They generally inhabit meadows, fields, and clearings in the east and north, but are most abundant on the southern prairies of central Canada.

Although chorus frogs belong to the tree frog family, their toe discs are minute and they rarely climb above the tops of tall grasses.

BULLFROG
RANA CATESBEIANA

This giant among North American frogs often attains a length of six inches or more. Bullfrogs occur naturally in lakes, rivers, and large ponds east of the Rockies from southern Canada to northern Mexico. They have been widely introduced into the western United States. Breeding occurs from late May to early July, when the deep "jug-o-rum" bellows of the males resound on warm evenings.

Bullfrogs are collected for dissection in school and university courses and their meaty hind legs are enjoyed as a delicacy by some gourmets.

RACER
COLUBER CONSTRICTOR

Racers are nervous, quick-moving, harmless snakes of open, bushy, or lightly wooded areas. They prey mainly on small rodents, frogs, lizards, and insects. Their scientific name was a mistake, for they never constrict, although they sometimes use a loop of their body to hold their prey down before swallowing it.

Ten varieties of racers occur in the United States. There are three Canadian racers, also widespread in the United States, which are uniformly colored above when adults, but have a pattern of dark blotches when newly hatched.

The Blue Racer is greenish or grayish blue above, and pale below. It is the largest of the racers, attaining a maximum length of about 70 inches.

This is the smallest of the three kinds of rattlesnakes that are Canada's only poisonous snakes (the United States has more). It rarely exceeds 30 inches in length. A pair of enlarged teeth at the front of its mouth are hollow, resembling short, curved hypodermic needles; through these, poison may be injected.

Because this poison is toxic enough to sometimes kill an adult, antivenin is on hand at hospitals within its range.

The Massasauga occurs from the southern Great Lakes region southwest to Texas and Arizona. It frequents low, swampy areas and feeds largely on frogs and mice.

MASSASAUGA RATTLESNAKE
SISTRURUS CATENATUS

COMMON GARTER SNAKE
THAMNOPHIS SIRTALIS

The Common Garter Snake occurs widely throughout the United States and Canada in a wide variety of color patterns. The specimen shown here, from eastern Ontario, is black with lemon-yellow stripes. Garter snakes are most usually from 24 to 32 inches in length, with the rare individual growing to over 53 inches.

They feed upon earthworms, frogs, fish, and occasionally mice. In winter they often den up in cracks or fissures in the earth, beneath the frost line, sometimes by the thousands. The young are born live, usually 20 to 40 in a litter, with the maximum recorded being 98. The flickering tongue is not, as some people suppose, a stinger of some sort, but is a sensory apparatus used to detect odors.

BLACK RAT SNAKE
ELAPHE OBSOLETA OBSOLETA

The Black Rat Snake may be the largest Canadian snake, and one of the largest in the United States. Its maximum recorded length is 8.2 feet, and individuals over five feet are not uncommon. It feeds largely on mice, rats, and some birds, and can quickly suffocate its prey by constricting it within tight coils of its body. An excellent and frequent climber, it generally inhabits woodlands and uplands away from water.

It is most commonly seen in the rocky, largely non-agricultural terrain of the Rideau Lakes region in eastern Ontario, southward to Tennessee and west to Oklahoma.

SPINY SOFTSHELL TURTLE
TRIONYX SPINIFERUS

The flat, almost circular upper shell of the Spiny Softshell Turtle resembles a pancake, and is covered by leathery skin rather than by the hard, horny scutes of most turtle shells. Large females may attain a shell length of 17 inches.

Softshells have been found in Canada in southern Ontario, southwestern Quebec, Lake Erie, and in the Ottawa, St. Lawrence, and Richelieu rivers, as well as through much of the central and southeastern United States, and as far west as Montana and southeastern California.

They are very aquatic and prefer areas of sand and mud. Here they often lie with only their nostrils above the shallow water. They eat crayfish, aquatic insects, and fish.

Rarely seen, they are savage in disposition if molested, and capable of inflicting serious injury with their razor-sharp jaws.

COMMON SNAPPING TURTLE
CHELYDRA SERPENTINA

The snapper is one of North America's largest freshwater turtles, attaining a shell length of 18 inches and weights exceeding 33 pounds. Its serpentine neck, massive head, muscular legs, and relatively long tail make it seem even larger.

Snappers are defensive and exceedingly ugly in disposition if confronted on land, but in the water they usually slip quietly away from any disturbance. They occur in ponds, lakes, rivers, and streams in eastern North America.

They consume a variety of aquatic plants and many kinds of animals, including fish, frogs, birds, and small mammals. They are effective scavengers, cleaning up dead fish or drowned animals. Their round eggs resemble ping-pong balls, and the young usually hatch in September or early October.

ATLANTIC RIDLEY
LEPIDOCHELYS KEMPI

This smallest of the sea turtles occurs from Canada to Mexico and across the Atlantic to Europe, but its only nesting beaches are on the Gulf of Mexico, mainly just north of Tampico where, in 1947, as many as 40,000 nesting females crowded the beach in a single day. Despite legal protection and military patrols of the beach at nesting times, today these numbers have dwindled to sometimes only a few hundred.

Nest robbing and the taking of females for food and leather caused much of the decline, and increasing numbers are being accidentally drowned in shrimp trawls. Protective enclosures for egg-hatching and attempts to establish nesting beaches in Texas may help the Atlantic Ridley survive.

ATLANTIC LOGGERHEAD TURTLE
CARETTA CARETTA CARETTA

The Atlantic Loggerhead Turtle is found in the waters off North America's east coast. Individuals may attain a carapace (shell) length of almost nine feet and weigh up to 1,000 pounds, although a weight of about 300 pounds is more usual.

In the open sea, these turtles spend much of their time floating idly on the surface of the water. They feed upon sponges, jellyfish, mussels, clams, oysters, shrimp, and a variety of fish.

Nesting takes place in temperate waters and is usually accomplished on open beaches by the female, who comes ashore at night and digs the nest in the sand by means of her flippers. The round, white, leathery eggs, as many as 126 in a clutch, are then covered over and the sand packed down on top of them. In a period of up to 68 days the eggs that have not fallen victim to predators hatch, and the young loggerheads struggle to the surface and make their way to the sea.

LEATHERBACK TURTLE
DERMOCHELYS CORIACEA

The largest living turtle, the Leatherback may attain a total length of over six feet, and a weight of over 1,000 pounds.

Unlike other marine turtles, the Leatherback lacks a scaled shell. Instead, it is covered with a layer of spongy, oily tissue, protected by a smooth leathery skin with prominent longitudinal ridges.

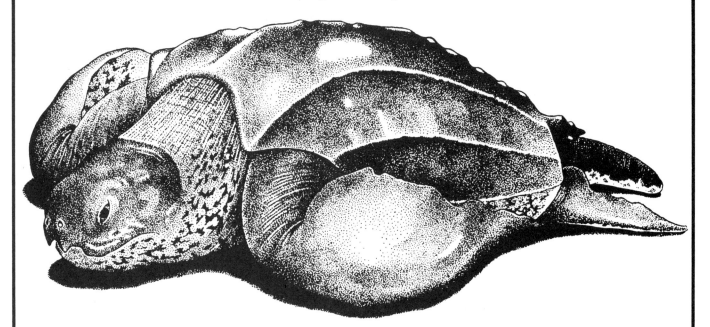

This sea turtle is global in distribution, unrecorded only in the Arctic and Antarctic. It nests on warm tropical beaches; in the Atlantic, nesting has not been recorded north of Florida. Only females come ashore, but they appear several times in one laying season to bury clutches of 60 to 160 eggs. The eggs hatch between 56 to 72 days later, and the young are vulnerable to many predators, both on their way to, and in, the water.

During the summer months, Leatherbacks have been observed as far north as central Labrador and southern Alaska. Because of their size, they can conserve body heat and maintain a temperature as much as 33°F greater than that of the water. In common with other marine turtles, this reptile's front legs have developed into large flippers to propel it. It feeds on jellyfish and other soft-bodied sea animals, often near the ocean's surface but sometimes on the bottom, as far down as 150 feet.

CROCODILE
CROCODYLUS

Only 21 kinds of crocodilians now represent this group, which flourished during the dinosaur era. True crocodiles are found in the warmer waters of Africa, Asia, Australia, and America in both fresh and salt water habitats. These reptiles differ from alligators and caimans by having an exposed lower tooth on each side when their jaws are closed. Their snouts are generally narrower than the blunt-snouted alligators and caimans, but broader than the very narrow-jawed gharials. Crocodiles vary in size: The American Crocodile may measure up to 23 feet in length, but the maximum length of adults of some other kinds is under four feet long.

Hunted extensively for their skins, large crocodiles are becoming scarce in the wild; world commerce in animal skins is regulated by the Convention on International Trade in Endangered Species. They feed upon fish, birds, and mammals—only rarely do they attack humans. They are fond of basking in the sun in groups with mouths agape. In some areas, certain birds wander freely among the basking creatures, picking leeches and parasites from their skins, and food fragments from their mouths.

AMERICAN ALLIGATOR
ALLIGATOR MISSISSIPPIENSIS

This reptile of the coastal marshes and inland waters of the southeastern United States was once considered an endangered species. Relentless hunting for hides reduced the numbers drastically in the 1960s until state and federal laws provided for complete legal protection. In addition, there are now import and export restrictions controlling trade in alligators and their skins.

Although poaching and illegal trade are always threatening, the American Alligator is recovering in the wild. They can be found from North Carolina to Texas. There are also thousands of them in zoos, where they breed successfully.

ALLIGATOR LIZARD
GERRHONOTUS COERULUS PRINCIPIS

The Alligator Lizard may have been named for its likeness to a miniature alligator, but its habitat and size bear no resemblance. It is, however, the largest lizard of the Pacific Northwest, attaining a total length of more than 8.5 inches.

It lives in dry, often rocky, wooded or partly wooded areas, sometimes in grassland, in the southern mainland of British Columbia, on Vancouver Island, and in the northwestern corner of the United States. Its most prominent characteristic is the longitudinal fold of small, granular scales along each side, which contrast with the relatively large scales of the back and belly, and which may allow expansion of the sides when the animal is breathing, feeding, or (in the case of the female) carrying young.

It is wary in the open and is usually found under cover of bark, logs, or stones. Insects and spiders are its main prey.

Section Two
BIRDS

These two birds come on land only for nesting, and spend the rest of the year offshore in favorable feeding waters. They nest in colonies on cliffs or among boulders and rocks, and lay a single egg, which is then incubated for about a month.

They are good swimmers and obtain their food by diving, often to depths of up to 65 feet. They occur mainly along the Atlantic coast and in some parts of the Arctic.

Some Common Murre colonies are enormous and may number nearly half a million birds.

RAZORBILL
ALCA TORDA

COMMON MURRE
URIA AALGE

NORTHERN GANNET
SULA BASSANUS

With a six-foot wingspan, the gannet is the largest seabird breeding in North American waters. It feeds primarily on surface-dwelling fish such as herring and mackerel, which it takes by diving from heights of up to 140 feet and plummeting into the water at great speed and with considerable force. The bird's skull is especially strong, and a system of air sacs also helps to absorb the shock of these plunges.

Gannets nest on the steep cliffs of Canada's east coast. The nests are large and are made from a mixture of vegetation, seaweed, feathers, and earth. These materials are sometimes cemented with guano.

North American gannets winter at sea from Virginia to southern Florida, and return to their breeding grounds in early April.

DOVEKIE
ALLE ALLE

These small, North Atlantic sea birds nest in crevices and clefts in high, rocky coasts, where they form densely populated breeding colonies. They are common today on the west coast of Greenland but are not known to nest in northern Canada or Alaska.

A tiny wing bone of a Dovekie from 40,000-year-old sea deposits near Cape Storm on southern Ellesmere Island may indicate that breeding colonies once existed there also. This insignificant-looking fossil is the first evidence of an ice age bird from the high Arctic of North America.

ATLANTIC PUFFIN
FRATERCULA ARCTICA

The Puffin is sometimes known as the "sea parrot" because of its distinctive bill, which in the nesting season takes on vivid colors of red, blue-gray, and ivory. The colors recede when the nesting season ends.

Winters are spent at sea in the North Atlantic, but no one is sure just where. The birds breed along the coastline of northeastern North America.

The largest North American colonies are found on islands off Newfoundland, one of which is occupied by an estimated 300,000 puffins.

The nest is actually a burrow which the bird digs into soft, turfy slopes for a distance of about three feet, and in which one egg is laid. The Puffin is an expert swimmer and often emerges from an underwater foray with six, eight, or more small fish dangling from its serrated bill. Puffins suffer considerable losses from Herring and other types of gulls, whose numbers are increasing because of human effluence.

GREATER PRAIRIE-CHICKEN
TYMPANUCHUS CUPIDO

The story of the Greater Prairie-Chicken is one of response to environmental change. A bird of the tall-grass prairies of midwestern North America, it moved into the prairie provinces of Canada in the 1880s, when a few wet years, agricultural settlement, fire suppression, and the disappearance of the bison produced ideal habitat. It became abundant and eventually spread eastward.

As the prairie gave way to intensified agriculture, so did the Greater Prairie-Chicken, and by the 1930s it was almost extinct. It now survives in scattered areas of southern Canada and the midwestern states. In the United States, land is being acquired for managed habitats to save the bird from extinction.

AMERICAN WHITE PELICAN
PELECANUS ERYTHRORHYNCHOS

The American White Pelican is a large, web-footed bird with an enormous throat pouch for scooping up fish. It is now found mainly in the western part of North America around inland lakes and marshes.

Pelicans are heavy-bodied birds with short legs and thick, rough plumage. They have wingspans of up to ten feet.

The American White Pelican catches fish by scooping them up in its pouch as it swims, while the brown pelican dives from the air to catch its prey, and is usually found along seacoasts.

Although pelicans are awkward-looking birds, they are very graceful in flight. They fly in a flock in a long line, with their elongated necks bent back over their bodies. They flap their wings only 1.3 times per second, while a Ruby-throated Hummingbird beats its wings 50 to 70 times per second.

The greatest cause of the American White Pelican's continuing slow decline, and its consequent designation as a threatened species in some areas of North America, has been human disturbance.

CANADA GOOSE
BRANTA CANADENSIS

This is probably the best-known goose in North America, where its strong, distinctive call is associated with the onset of spring and fall.

They are prized by hunters as one of the choicest game fowl, but their wariness, together with protective regulations, prevent any serious depletion of their numbers.

These birds mate for life and the family group remains together for several months after the hatching of the young. A gander protecting the nest makes a very formidable adversary, and his wings are capable of delivering a blow of surprising force, sufficient to rout foxes and similar predators, including overly curious humans.

TRUMPETER SWAN
CYGNUS BUCCINATOR

The Trumpeter Swan, largest and rarest of the world's eight swan species, was once a common nesting bird in north, west, and central North America. It was hunted extensively by natives for food and feathers, and its numbers began to decline when a market developed in European settlements for its skin, feathers, down, and quills. The decline continued with the gradual loss of nesting, feeding, and wintering habitats, especially in the United States, due to expanded land use. By the early 1900s the bird's extinction was thought near.

Now legally protected in Canada and the United States and provided with sanctuaries, its numbers have slowly increased through emergency winter feeding, habitat restoration, and controlled relocation of populations. Trumpeter Swans presently number more than 5,000.

Although still carefully monitored, they have been removed from the list of endangered species.

WHOOPING CRANE
GRUS AMERICANA

Never very abundant, the Whooping Crane suffered in the late 1800s from indiscriminate shooting, habitat disturbance, and the draining of the large, isolated marshes that it frequented.

In 1941 there were only 21 wild birds and two captives. Still on the endangered list and carefully monitored, the cranes now total 107 in two free flocks, plus 37 in captivity.

Total legal protection, public interest, and protected breeding grounds in Canada and wintering grounds in the United States, as well as artificial incubation, foster parenting by Sandhill Cranes, and the establishment of an additional flock breeding in Idaho, have all helped in rescuing the Whooping Crane from probable extinction.

GREAT BLUE HERON
ARDEA HERODIAS

This stately bird is the largest and most widely distributed of the North American herons, and is found in many areas of Canada and the United States—including the Florida Keys.

Sociable by nature, it lives in colonies, building nests in the uppermost branches of tall trees. When necessary, however, it will nest in smaller trees or even on the ground.

Its principal food is fish, which it takes by wading in shallow water. Frogs, eels, meadow mice, snakes, and rats are also consumed, and this bird is also adept at catching insects.

Very few other birds or animals risk encounters with the Great Blue Heron, for it is courageous and has a sharp, powerful bill capable of inflicting serious wounds.

Losses occur, however, when unattended eggs are taken by crows or ravens, or when nesting birds are disturbed by humans.

CONDOR
VULTUR GRYPHUS

To see a South American condor gliding in for a landing on the face of a cliff is a splendid sight. No airplane is as maneuverable as this giant bird of prey as it soars gracefully on motionless wings. At the right moment, it drops its landing gear (its feet), brakes with its wings, raises and spreads its tail, and then lands.

This fierce-looking bird belongs to the same order as the vulture. There are ten species, all found in North or South America. All have ugly, naked heads, large hooked beaks, and huge wings.

The largest species is the black, gray, and white Andean condor: It is over four feet long, with a wingspan of about ten feet. It usually lives in the Peruvian and Chilean Andes at heights of 10,000 to 18,000 feet, but is also found on the rocky seacoast. It is often seen soaring in search of carrion, but its diet also includes young goats and lambs. The California condor is the last descendant of a giant scavenger that lived in the United States during the Pleistocene Age. Hunted extensively by ranchers, only about 60 of these birds survive today.

This small owl is a bird of the treeless, short-grass country of western North America, from southern Canada into Mexico. It also occurs in Florida, the West Indies, and Central and South America.

The Burrowing Owl depends mainly on the abandoned burrows of prairie dogs and, to some extent, upon those of badgers, ground squirrels, woodchucks, wolves, foxes, skunks, and armadillos for nesting sites and shelter.

Although it is presently not rare, the Burrowing Owl could become limited in numbers because of its preference for uncultivated land and its dependence upon other animals for nesting sites.

BURROWING OWL
ATHENE CINICULARIA

SNOWY OWL
NYCTEA SCANDIACA

The Snowy Owl is a large, diurnal bird of prey of the Arctic regions of the world. In North America it nests north of the tree line. Its life cycle is intimately dependent on an abundance of lemmings, which form an important part of its food supply.

In years when food is plentiful the Snowy Owl reproduces; in years of scarcity, nesting does not take place and the birds wander off to the south, and then are sometimes seen in the northern regions of the United States. These cycles occur every four or five years.

The Peregrine Falcon, once numerous throughout North America, could have become extinct because of the effects of chemical pesticides such as DDT. However, public awareness and conservation programs seem to have saved the species and it now seems to be occupying its former range again. During the past 25 years, however, the Peregrine has suffered a dramatic population decline, and is presently on the list of endangered species in both Canada and the United States.

This bird is one of nature's swiftest fliers, attaining dive speeds of over 200 miles per hour. Its prey consists primarily of other birds. From them, it ingests the chemicals that have already led to its extinction in some regions.

PEREGRINE FALCON
FALCO PEREGRINUS

GOLDEN EAGLE
AQUILA CHRYSAETOS

This bird's range extends throughout most of the northern hemisphere.

A splendid flier, the Golden Eagle reaches speeds of 150 to 200 miles per hour when diving. Most of its prey is taken on the ground and consists of mammals such as foxes, rabbits, hares, and so on. Large birds, such as geese and cranes, are occasionally struck in mid-air. The eagle does not, contrary to legend, prey to any extent on domestic animals.

Nests are built on crags or in trees, and are sometimes occupied for generations. The young hatch at intervals of several days, and in many cases where two young eagles hatch, the elder may kill the younger.

BALD EAGLE
HALIAEETUS LEUCOCEPHALUS

This majestic bird, the national symbol of the United States, is distributed across much of North America but is most common in the Pacific Northwest and Alaska. The adult can be distinguished by its white head and neck, which takes about four years to develop.

Adult males have body lengths of 30 to 34 inches and wingspreads of 72 to 85 inches. Females are slightly larger.

The Bald Eagle is a scavenger rather than a predator, subsisting mainly on fish. It is most commonly found along seacoasts where cast-ups provide most of its food. It occasionally makes its own kills. When fish are not available it may take a few birds.

Bald Eagle numbers have declined, but fortunately, populations appear to be holding their own in recent years.

PIPING PLOVER
CHARADRIUS MELODUS

"Unspoiled, undisturbed, clean, sandy beaches on seashores and inland lakes" describes ideal vacation sites. They are also ideal nesting grounds for the Piping Plover, whose breeding range extends from Virginia to the Maritimes and from the Great Lakes to Alberta and Nebraska.

Once considered a game bird and exposed to many years of spring and fall shooting, the plover had declined drastically by the early 1900s. Legal protection as a migratory non-game bird allowed it to make a recovery, but in recent years, with the increased recreational use of beaches, the plover's survival, particularly in eastern North America, is again threatened.

In summer, this colorful bird with a red cap occurs in most of the forested regions of northern North America. The Sapsucker, together with the Flicker, is one of the few species of migratory woodpeckers that occur in this region. It winters from the southern United States south through Central America. Males have a very distinct red patch on the throat, while the same area in females is white. They feed, to a great extent, on sap and cambium (the inner bark), which they obtain by drilling holes in the trunks of trees. This peculiar pattern of holes indicates their presence (see inset).

Insects are also part of this bird's diet and are captured, as is the sap, by means of a long bushy tongue.

YELLOW-BELLIED SAPSUCKER
SPHYRAPICUS VARIUS

COMMON NIGHTHAWK
CHORDEILES MINOR

The Common Nighthawk occurs in southern Canada, and its range extends south to southern Louisiana, California, Nevada, and Arizona. It is often seen flying high over city streets and rooftops, or in open country areas on summer evenings.

A hollow, booming sound is produced by the wings as the bird pulls out of a steep dive while in pursuit of flying insects.

Nighthawks catch mosquitoes, flying ants, and other flying insects. Their enormous mouths, with their surrounding bristles, are ideally suited for aerial capture.

Nighthawks build no nests, but lay their eggs on the ground or, in cities, on flat gravel rooftops.

Look and listen for nighthawks over the city some evening in the summer.

NORTHERN SHRIKE
LANIUS EXCUBITOR

The Northern Shrike is known also as the "butcher bird" because of its unusual practice of impaling prey on thorns, barbed wire, and so on, much in the way butchers hang up meat in their shops. Mice, other birds, and large insects form the bulk of the shrike's diet.

This is a strong passerine bird of medium size that has evolved characteristics of typical birds of prey—such as the strongly notched bill to tear flesh. Its plumage is a soft silver-gray, combined with dark brown, almost black, wings and tail. Breeding in northern Canada and Alaska, it winters in southern Canada and the northern United States. It has a well-developed song and is capable of imitating the songs of several other birds.

SNOW BUNTING
PLECTROPHENAX NIVALIS

The Snow Bunting breeds across the Arctic regions of the world. Its appearance in the southernmost parts of Canada heralds the approach of winter. They are usually seen in small flocks, along shorelines, and in open fields, where they feed mainly on weed seeds and insects. A sizeable flock swirling over a field in winter can be a spectacular sight.

It usually departs for Arctic breeding grounds in early spring, where it raises four to seven young.

BARN SWALLOW
HIRUNDO RUSTICA

These graceful little birds may have been named for their affinity for the old wooden barns once found on many farms, which they could enter and leave with ease through open doors, chinks in the walls, and so on.

Swallows are sociable by nature and often gather in large flocks of different species. They spend a great deal of their time in the air and consequently nearly all of their food is captured while they are on the wing. They are a familiar sight in rural areas, following the farmer as he plows, and catching the insects stirred up by his progress.

Both male and female share in the construction of the nest, which sometimes takes eight 14-hour days to complete. The nests are usually constructed of mud mixed with straw, grasses, or horsehair, and are cemented to the vertical surfaces of old beams or rafters.

RUBY-THROATED HUMMINGBIRD
ARCHILOCHUS COLUBRIS

The Ruby-throated Hummingbird is only one of the numerous beautiful hummingbird species found in North America. It occurs all across the United States and in southern Canada.

The hummingbird family contains the smallest of birds and they are so named because of the characteristic humming sound made by their tiny wings. They are capable of rapid forward flight, of remaining stationary in mid-air while feeding at flowers, and even of backward flight for short distances.

Hummingbirds feed on minute insects and flower nectar, which they obtain by means of a long, extensile tongue. They are found in gardens, orchards, and woodland clearings where flowers, either wild or cultivated, are present to provide nourishment.

Because of their size, hummingbirds are often mistaken for large hawk moths, which also feed at flowers.

AMERICAN CROW
CORVUS BRACHYRHYNCHOS

The loud and frequent "caw-caw" of this good-sized bird, with its shiny black plumage, is quite familiar. Their return in late winter means the advent of spring to many people in Canada and the northern United States.

The Common Crow, in spite of its bad reputation as a predator, is a useful bird because it consumes large quantities of harmful insects.

The crow nests in trees, where four to six eggs are laid. After the breeding season, immatures and adults often assemble in large flocks on their way to the southern wintering grounds. Many birds remain year-round in suitable areas.

COMMON RAVEN
CORVUS CORAX

Aggressive, clever, and inquisitive, this large black bird is well distributed in the northern hemisphere. Its habitat and that of man do not overlap to a great extent.

The raven is omnivorous but tends to be predatory. Its diet embraces a wide variety of animal and vegetable matter.

One can distinguish it from the American Crow by its larger size, its voice—a peculiar "crâaw"—and its wedge-shaped tail.

ESKIMO CURLEW
NUMENIUS BOREALIS

Now almost extinct, this shorebird once migrated in huge flocks between the remote breeding grounds of the taiga and wintering grounds in Argentina. Each year, on its way there and back, it faced a slaughter in fall, winter, and spring by gunners, who sometimes filled wagons with the birds.

Early in the 1900s the Eskimo Curlew was considered extinct, but a few sightings in the past 25 years in Canada and the United States indicate the existence of a small breeding population.

The breeding grounds are yet to be discovered in the north. The bird now has complete legal protection.

AMERICAN ROBIN
TURDUS MIGRATORIUS

One of the most widely accepted signals that spring has arrived once again is the return, usually in March in northern North America, of the robins from their southern wintering grounds. Named by the early English settlers because of its color resemblance to the English Robin, the American Robin is actually a thrush, one of the largest in North America.

KIRTLAND'S WARBLER
DENDROICA KIRTLANDII

This bird is known to nest only in the United States and to winter in the Bahamas. Only on the ground, in the dense undergrowth of young jack-pine forests in central Michigan, does it find its required nesting conditions. These appear about ten years after a forest fire and remain for only a few years, until the new trees shade out the undergrowth.

Such exacting requirements, for fewer than 500 of the birds, prompted the cooperation of federal, state and private agencies, with the help of public opinion, in setting aside large reserves, totalling thousands of acres, in jack-pine country. Here, management through controlled burning will ensure successions of young forests and the continued existence of nesting grounds for this endangered bird.

Section Three
MAMMALS

EASTERN CHIPMUNK
TAMIAS STRIATUS

This interesting little inhabitant of northeastern forests and fence-rows in the United States and Canada is a favorite of young and old alike. Cottagers and campers have found that, if unmolested, it soon becomes bold enough to accept food held out to it, much of which is hoarded for the future.

Chipmunks construct extensive burrow systems, often more than 11 feet in length and with one or more well-concealed entrances. In addition to the main chamber, blind storage tunnels are constructed to accommodate the winter food supply. The sleeping quarters are kept scrupulously clean—shells, husks and feces are stuffed away into refuse tunnels.

Chipmunks eat a wide variety of seeds, fruit, and nuts. They are also fond of corn and sunflower seeds. In autumn, they may store as much as seven liters of food for winter use. The scientific name Tamias means "steward" in Latin and seems to be most appropriate.

They are preyed upon by hawks, foxes, and weasels, and sometimes also fall victim to domestic cats.

NORTHERN FLYING SQUIRREL
GLAUCOMYS SABRINUS

These appealing little creatures are found throughout most forested areas of the United States and Canada, but are rarely seen because of their nocturnal habits. A furry membrane uniting the front and back legs enables them to glide from higher branches to lower ones with great dexterity. This is a sociable little animal with strong maternal instincts.

Their large, dark eyes give flying squirrels a look of lively intelligence, and they have indeed been known to make affectionate and playful pets. The males and females frequently occupy separate nests in the summer, but in the winter months they often live in "communes" in hollow trees or stumps. Their fondness for maple sap has often led to their tumbling into sap containers in the spring and drowning. Their principal enemies are owls and cats.

RICHARDSON'S GROUND SQUIRREL
SPERMOPHILUS RICHARDSONII

This squirrel inhabits the grasslands of the Great Plains. It can easily be identified by its frequently assumed upright posture and quick, quivering tail, which is not long and bushy like that of the tree squirrel.

It lives in burrows and feeds on a wide variety of plants and their seeds. Non-vegetarian fare is not rejected either.

During winter, when the exposed plains turn into a harsh and deadly environment, this plump-bodied little animal hibernates securely in the depths of its burrows.

RED SQUIRREL
TAMIASCIURUS HUDSONICUS

Wide distribution and diurnal habits make this active little squirrel one of our most frequently seen wild mammals. Its loud, rolling "tcher-r-r" and scolding chatter are familiar sounds in the northern forests.

The red squirrel constructs its nest in the branches or cavities of trees or stumps. In winter it lives on food diligently stored up during late summer or fall. A single cache may contain several hundred spruce or pine cones.

Familiar to almost everyone, found in every state and province, this small rodent came to North America in the baggage and stores of early settlers. In their albino form they are seen as laboratory animals and pets.

Much like the rat, the house mouse breeds throughout the year, the gestation period lasting 21 days, producing an average litter of between four and seven. Several litters are born each year. Sexual maturity occurs at 35 days of age.

They are chiefly nocturnal in activity. Their food consists of grains, fruit and vegetables, stored food, and refuse.

The predations of cats, dogs, snakes, hawks, owls, weasels, opossums, skunks, raccoons, and foxes help to control their numbers.

HOUSE MOUSE
MUS MUSCULUS

KANGAROO RAT
DIPODOMYS ORDII

This animal measures up to 12 inches in length, including its long tail. It occupies the Great Plains, the Great Basin, and the Mexican highlands of western North America, from central Mexico north to Canada. In Canada it is found only in the Great Sand Hills area of southwestern Saskatchewan and adjacent areas of Alberta. Their abodes are shallow burrows dug in loose sand.

In the wild they are solitary and quite belligerent by nature, frequently engaging in combat by leaping into the air and slashing at each other with powerful hind feet.

Against more deadly enemies, such as rattlesnakes, they often turn their backs to the foe and vigorously kick sand into its face.

When fleeing from danger they use their powerful hind legs to cover the ground in a series of 6.5-foot-long hops, in the manner of a kangaroo.

BROWN RAT
RATTUS NORVEGICUS

Originating in Asia and arriving in North America about 1775, the fierce and aggressive brown rat is man's worst animal pest. Neither man nor natural predators have made any serious inroads in its population.

Averaging about 16 inches in length, these animals live in organized social colonies, usually dominated by the larger males. Active at all times of the day, they do enormous damage each year, not only in food destroyed and polluted, but also in the destruction of electric and telephone cables, water pipes, furniture, and so on.

They frequently kill chickens, ducks, and even lambs and piglets. They are known carriers of several deadly diseases, one of which is bubonic plague.

The first step in controlling these creatures is good sanitation and the proper disposal of garbage, together with the rat-proofing of food storage areas.

VANCOUVER ISLAND MARMOT
MARMOTA VANCOUVERENSIS

The world population of this animal, which is about the size of a woodchuck, is found on Vancouver Island, and there only on a few of the higher mountains. Its very restrictive habitat is alpine and sub-alpine meadows and avalanche slopes. It has suffered from human interference through logging, recreation and hunting. Its numbers have been declining since its discovery in 1910 and it is estimated that fewer than 70 animals now remain.

This endangered species has gained the attention of naturalist groups and government agencies. Protective measures call for no further habitat disturbances, especially logging, mining, and/or developments; complete protection from hunting; and a ban on scientific collecting.

This aquatic rodent is found across most of North America with the exception of the Arctic tundra. Fully grown, it measures about 24 inches in overall length and weighs about 3.3 pounds. As would be expected, it is an excellent swimmer, capable of traveling 300 feet under water and staying submerged, when necessary, for more than fifteen minutes at a time.

Their houses are constructed of bullrushes, weeds and packed mud, with separate sleeping platforms for each member of the family. They also build dens in stream banks, with the entrances under water. Very quarrelsome even among themselves, they are vicious fighters and have been known to attack humans without provocation.

In addition to a wide variety of vegetation, they also feed on freshwater mussels and are reported to eat frogs and small turtles. Their chief predator is the mink, but while on land they also fall prey to foxes, coyotes, and lynx, as well as some of the larger avian predators.

MUSKRAT
ONDATRA ZIBETHICUS

BEAVER
CASTOR CANADENSIS

The beaver, Canada's national symbol, was also the country's first natural resource to be exploited, and at one time beaver pelts were the unit of currency in the new land. The beaver population was almost wiped out by 1930, but conservation measures since then have restored their numbers to relatively healthy levels in both the United States and Canada.

Beavers have a well-developed social hierarchy in which the family is the basic unit, and the female the central figure in each family. The usual family group consists of the adults, the kits, and the yearlings of the previous year, bringing the average family group to 10 or 12 individuals. Adult weight varies from 33 to 77 pounds, with the average being about 44 pounds.

The bear, wolf, coyote, fisher, wolverine, otter, and lynx prey upon the beaver, who is, nevertheless, a powerful antagonist when at bay. Beaver lodges, made of tangled sticks and caked mud, offer protection that even black bears have difficulty in breaking through.

Beaver dams are usually about 165 feet in length, over six feet wide, and about ten feet across the base. Their dams help to maintain water levels in forest streams, thus providing habitat for fish and waterfowl.

AMERICAN PIKA
OCHOTONA PRINCEPS

This stocky, tailless little animal with the Roman nose is about 6.8 inches long and is found in the mountain regions of western North America. Diurnal in habits, it appears in early morning and disappears into a grass-lined nest, hidden in the rock crevices, shortly after sunset.

The pika spends considerable time sunning itself on a favorite lookout rock, against which its salt-and-pepper coat is difficult to distinguish.

Remaining active throughout the winter under the snow, the pika subsists on stacks of cured grasses and sedges, which it stockpiles during the summer months.

Preyed upon by eagles and hawks, as well as bears and foxes, its most dangerous foe is the ermine, which is capable of following it into its rocky tunnels.

The hoary marmot, so called because of the mantle of white fur that covers its shoulders and back, is well known to hikers in the western mountains of North America. A high-pitched whistle often welcomes visitors to the alpine country and warns the members of the colony of approaching danger.

Fattened by a summer of feeding on lush alpine plants, marmots, like some other members of the squirrel family, spend the winter months in hibernation far beneath the snow.

These animals, sometimes called "whistlers," are slightly larger than their relative the woodchuck, and can weigh up to 30 pounds.

In early summer marmots spend hours engaged in playful wrestling matches. The pushing and grappling can be quite vigorous, but marmots, like Olympic wrestlers, have only to give the right signal—in this case a sharp yelp—to end the bout and prevent injury.

HOARY MARMOT
MARMOTA CALIGATA

ARCTIC HARE
LEPUS ARCTICUS

The arctic hare inhabits the tundra regions of Canada from Newfoundland west to the Mackenzie delta, and north to the tip of Ellesmere Island.

On the High Arctic islands where hares retain their white coats all year, they sometimes band together into groups of up to 200 individuals.

As adults they have few enemies besides the wolf, though the young hares are taken by Gyrfalcons, Snowy Owls, arctic foxes, and ermine.

In the extreme cold of the Arctic winter, hares dig dens in hardened drifts. While resting they sit on their well-furred hind feet, hunched into a heat-conserving ball.

When alarmed, however, they rise up on their hind legs to look for danger, then bound off, hopping upright on their hind legs like kangaroos, at speeds estimated to be in excess of 30 miles per hour.

AMERICAN PORCUPINE
ERETHIZON DORSATUM

This slow-moving creature is North America's largest rodent, next to the beaver, and is distributed throughout most of the timbered areas of the United States and Canada. Large males may reach 35 inches total length and weigh up to 26 pounds. The head, neck, and rump are protected by quills, the tips of which are barbed; this makes removal from the victim both painful and difficult.

When under attack the porcupine presents its posterior to the adversary and lashes out with its spiny tail. The quills are so lightly fixed to the porcupine's body that they are easily detached and left imbedded in the attacker. Bobcats, wolverines, and fishers are the most adept at killing porcupines, as they have learned how to flip the animal on its back to expose its undefended underside.

Primarily nocturnal, porcupines are active all year. Their summer diet consists of a variety of shrub and tree leaves. In winter they feed on the cambium layer and inner bark of trees. Their fondness for salt often leads them to roadways where salt has been sprinkled to melt the snow. Around campsites they will gnaw on anything containing salt, such as canoe paddles, axe handles, saddles, etc. The young are able to move about quite briskly shortly after birth, and, unlike their stolid parents, are quite playful. Porcupines are excellent swimmers.

STAR-NOSED MOLE
CONDYLURA CRISTATA

Like others of its kind, the star-nosed mole is very powerful in relation to its size. Using its paddle-shaped hands together with a twisting motion of its compact body, it constructs a network of tunnels down to a foot or two beneath the ground, but above the water table. Its nests are made of dried grass and other vegetation.

Found over most of eastern North America as far north as James Bay, this animal is about eight inches in length, approximately one third of which is tail. The tip of the snout is expanded into a naked pink disc that supports 22 fingerlike tentacles or feelers, which give this creature its name.

Active for 12 months of the year, it spends a good deal of its time in the water where it is an able swimmer, and it has often been observed swimming under the ice in winter. The star-nosed mole prefers wet, swampy ground and subsists on a diet of worms, insects, and crustacea. It has few natural enemies but sometimes falls prey to the Great Horned Owl and to some of the larger fishes.

HAIRY-TAILED MOLE
PARASCALOPS BREWERI

About six inches long, with a short tail covered with stiff, black hairs, this animal is found in southwestern New Brunswick, through southern Quebec and Ontario, southward to North Carolina.

It prefers forested areas or old pasture land, where the soil is relatively dry and loose. Clumsy on the surface because of its short legs and rotated forefeet, the animal is capable of tunneling through loose soil at considerable speed, using its forefeet as earthmovers.

Permanent tunnels, together with the nests, are usually about 10 to 18 inches beneath the surface, but deeper during the winter months. This animal's minute eyes, which are generally hidden behind facial fur, can be protruded under conditions of pain or alarm. Its diet consists of insects, larvae and earthworms. This mole is preyed upon by owls, foxes, and the larger snakes.

This stocky, muscular member of the weasel family is distributed throughout most of North America. It prefers life on the shores of deep, clear water in rivers, lakes, large marshes, and ocean bays. It attains a maximum length of about 4.5 feet and weight of about 30 pounds and has the thick, lustrous fur characteristic of amphibious mammals.

Although wary in the wild, it is by nature a sociable, docile, playful animal, and is easily domesticated. The Old World species was once used to catch fish for the table.

Primarily nocturnal, otters remain active all year. Their diet consists mainly of fish, but they also eat insects, frogs, and occasionally small mammals such as muskrats.

Although they sometimes fall prey on land to wolves and coyotes, the otter's principal danger is from man. Between 15,000 and 20,000 of these animals die in traps each year in Canada alone.

OTTER
LONTRA CANADENSIS

SEA OTTER
ENHYDRA LUTRIS

Highly prized in the fur trade, the soft, thick, lustrous coat of this marine mammal caused its near extinction. It once ranged the coastal waters of the Pacific Ocean and the Bering Sea from northern Asia and the Aleutian Islands down to southern California.

By 1911 it was close to extinction, but an international treaty gave it complete protection. The world's sea otter population is now estimated at 25,000 to 40,000 individuals.

Recent transplants have re-established the species along parts of the Pacific coast, including the coast of British Columbia.

Oil pollution will always be a threat to the sea otter, which lacks deposits of fat under its skin and depends upon air trapped in its thick fur for insulation against the cold.

This aquatic member of the weasel family is found throughout most of northern North America and is our most valuable furbearing animal. Large males may reach a total length of two feet. As adults they are bold, ferocious, and virtually untamable, but if captured as kittens they are playful and can become attached to the person who cares for them.

The mink is a skillful hunter and preys on a wide variety of game, including muskrats, meadow voles, and cottontails, as well as fish, crayfish, and frogs.

The mink's principal enemies are Great Horned Owls, bobcats, wolves, and coyotes.

More than 72,000 wild mink were trapped for their fur during the 1971–72 fur season, but many more were raised on mink ranches for the same purpose.

A solitary animal, and mainly nocturnal, the mink is active all year.

AMERICAN MINK
MUSTELA VISON

LEAST WEASEL
MUSTELA NIVALIS

Occupying most of northern North America, this fierce little carnivore is scarcely larger than the mice on which it preys. Its total body length seldom exceeds eight inches.

Incredibly quick and agile, these tiny predators are seldom trapped except by accident. The coat, which is brown above with a white underside, turns completely white in winter throughout most of its North American range.

Living almost exclusively on mice, shrews, and lemmings, the least weasel also uses the fur of these animals to line its underground nests. The Eskimos in particular look upon this small hunter with great respect, and the capture of one is considered a good omen.

ERMINE
MUSTELA ERMINEA

This little carnivore, scarcely a foot long, is found throughout all of Canada, Alaska, and the northern United States. In summer its coat is a rich chocolate brown, except for the undersides of the body and legs. In winter the color changes to a clear white, broken only by an area of black at the tip of the tail. Muscular, agile, and curious, it has been known to clamber up a man's clothing to reach food, or to attack someone who releases it from a trap. It is occasionally seen during daylight hours but is primarily nocturnal.

Some ermine appropriate the burrows of mice or ground squirrels and adapt them for their own use. Others build dens in hollow logs, under tree roots, or in abandoned buildings.

The ermine's diet consists primarily of mice, but it also preys upon cottontails, and small hares, porcupines, squirrels, pikas, and rats. In turn it is hunted by coyotes, badgers, foxes, owls, and wolverines. The small skins are used in the fur trade for coat trimming, stoles, and neckpieces.

Last noted in Saskatchewan in 1937, this weasel is now considered extinct in Canada. It has fared little better in the United States. An animal of North America's arid, short-grass prairies, it lives primarily with, and on, prairie dogs.

Wide-scale poisoning programs to eradicate prairie dogs and the destruction of their grassland habitat have affected the ferret.

It now survives in a few places in the western United States where ranchers are compensated for not molesting prairie-dog towns and where management programs for the prairie dog and the ferret are being developed.

BLACK-FOOTED FERRET
MUSTELA NIGRIPES

AMERICAN MARTEN
MARTES AMERICANA

This member of the weasel family has long, lustrous fur. It was once found in a broad belt across the forested portion of northern North America, but excessive trapping and destruction of its habitat have depleted its numbers, and today it survives only in scattered pockets of land. Its insatiable curiosity and appetite, often mistaken for tameness, make it a rather easy victim for all sorts of traps.

The males are solitary and pugnacious, associating with the females only in the mating season in July and August. The young are born and raised in grass-lined nests in hollow trees, or in cavities in rocks.

Considered to be tree dwellers, martens in fact spend considerable time on the ground and are excellent swimmers, even under water. Their principal source of food is mice, but squirrels, snowshoe hares, and pikas are also popular. So are a variety of fruits and berries, insects, and some carrion.

AMERICAN BADGER
TAXIDEA TAXUS

Badgers are relatively large members of the weasel family, the American badgers being slightly smaller than those found in the Old World. They are equipped with a belly gland that emits a musky odor when the animal is excited. Large males may weigh up to 25 pounds.

Badgers are not very agile, and run close to the ground with a trotting movement when pursued. Their normal gait is a leisurely waddle. These animals live in burrows that may be as long as 30 feet and up to 10 feet deep, with grass-lined sleeping chambers at the end. In their search for food—mostly burrowing rodents—badgers tear up large areas of earth with the powerful digging claws on their forefeet.

A courageous and powerful fighter, the badger has few natural predators besides man. Once common on the prairies, its numbers are now greatly reduced. It is found from the mountain valleys of southeastern British Columbia south to central Mexico.

WOLVERINE
GULO GULO

One of the larger species in the weasel family, this stocky, muscular animal is today found in its North American distribution chiefly in the northern regions of Canada, between the tree line and the Arctic coast.

Pugnacious, bold, and curious like other weasels, the wolverine is omnivorous, consuming a wide range of roots and berries, small game, and fish. It has been known to kill animals as large as caribou and mountain goats.

The wolverine's range is extensive, individual animals having been trailed for 60 to 80 miles over the snow. Primarily solitary, they follow migrating herds of caribou and clean up carcasses left by wolves and bears, crushing the bones with their powerful jaws.

Averaging about 30 pounds in weight, the wolverine has been credited with the ability to defend its food against wolves and even grizzly bears. It is active both day and night, seldom seeking shelter even in the severest winter weather.

Skunks are well known to most people, usually by reputation rather than by first-hand experience. The animal's best-known feature is, of course, its ability to discharge twin streams of musky fluid from its anal orifice, which it does with uncanny accuracy when it feels itself threatened.

The great confidence it feels in its defensive equipment is no doubt partially responsible for its somewhat placid and sluggish nature.

Skunks are primarily nocturnal and have few natural enemies, one being the Great Horned Owl.

Other carnivores avoid them unless desperate for food.

There is a popular notion that skunks cannot eject their musk if they are lifted off the ground by the tail. This is not true.

STRIPED SKUNK
MEPHITIS MEPHITIS

RACCOON
PROCYON LOTOR

This highly intelligent animal occurs throughout most of the more southerly areas of Canada southward through to Central America. The average weight of an adult male is about 19 pounds; the largest specimen reported weighed over 62 pounds.

They live primarily in forested areas near water, where they can find some of their favorite food items such as crayfish, frogs, small fish, and turtles. They also eat wild berries, corn, meadow voles, and mice. Mainly nocturnal, they have few natural predators, although a few are taken by foxes, bobcats and coyotes. Generally mild-mannered and easily tamed, they will fight bravely against great odds if cornered.

RED FOX
VULPES VULPES

The red fox is found from coast to coast in North America. Resembling a small, slender dog, it is a shy, nervous animal and, for the most part, tries to remain hidden from view. Unlike the dog, it is equipped with furred footpads.

Foxes dig their dens in sandy or gravelly soil, and these may be anywhere from 10 to 30 feet in length, and may have two or three entrances. They are conscientious parents, and remain together as family units until autumn, when the pups disperse. Their winter diet consists mainly of small mammals: mice, squirrels, rabbits, and so on. In the summer months insects, crayfish, and vegetable matter play a more important part in their feeding habits.

The red fox, in turn, is preyed upon chiefly by coyotes and bobcats.

COYOTE
CANIS LATRANS

About the size of a small collie, this intelligent, social animal is found in Alaska, Canada, and the United States. The coyote feeds mainly on small mammals such as mice and hares, and on carrion and some vegetation. Packs will occasionally try their luck with deer, but single coyotes have little success in this area. They, in turn, are preyed upon by wolves, cougars, and bears.

Man has employed a variety of devices against the coyote, including bounties and strychnine. But coyotes destroy a number of agricultural pests, and so some second thoughts are now being held. It is to be hoped this animal will be permitted to survive, at least in wilderness areas.

WOLF
CANIS LUPUS

The wolf has been associated with the northern races of man for centuries and remains, even today, a creature of folklore and superstition. The coloration of wolves varies greatly, from snow white to coal black and all intermediate degrees of cream, gray, and brown.

A large male wolf may measure over six feet in total length and stand almost three feet high at the shoulders. Their weight can vary from 57 to 176 pounds. Wolves are found across most of North America except in the southeastern United States.

These animals have a well-developed social hierarchy. The pack leader is usually the largest and strongest wolf, followed by younger or senile males, then the leader's mate, then other females, and finally the pups in order of strength. All members of the pack accept responsibility for the care and training of the young, even to the extent of "baby-sitting" if both parents are away hunting. There have been very few authentic cases of wolf attacks on humans in North America. In those rare instances, it is believed that mistaken identity was the cause.

ARCTIC WOLF
CANIS LUPUS ARCTOS

This large wolf, which is more lightly colored than southern wolves, ranges year-round over most of the islands in the Arctic.

Roaming singly or in small packs, wolves hunt cooperatively, using strategy to outwit the swift caribou—their major prey species.

A large wolf can bring down and kill an adult caribou with a single crushing bite to the neck.

Survival of pups depends on food supply and many die young. Thus wolf numbers are adjusted to fit the availability of prey.

Though wolves were once shot on sight at Arctic bases, now only the Inuit hunters kill them. The furs are sold or used for parka trim.

LYNX
LYNX LYNX

An inhabitant of the forested areas from the northern United States north to the tree line, this member of the cat family is distinguished by a short body, long legs, large padded feet, and pointed ears tipped with long, black hair. It is a nocturnal, solitary hunter, first appearing shortly before dark and remaining active until shortly after sunrise.

The lynx is an excellent climber and occasionally swims across rivers and streams. Its major food item is the snowshoe hare, but it sometimes takes ducks, foxes, and skunks, and has been known to kill deer. Its main enemies, besides man, are cougars and wolves.

The soft, lustrous fur has been a popular fashion item at various times in the past and has led to the demise of many thousands of these animals.

BOBCAT
LYNX RUFUS

Similar in appearance to the lynx, the bobcat is found in all parts of the United States except the midwestern corn belt, throughout Mexico, and in the southernmost parts of Canada. Generally smaller than the lynx, the bobcat also has a more heavily spotted coat. Some individual bobcats have, however, attained weights of almost 70 pounds.

A bold, solitary stalker of small game, the bobcat's diet consists mainly of mice and rabbits, but it has been known to kill muskrats, mountain beavers, foxes, weasels, and so on. It will occasionally attack deer, but is generally too small to have much success in these ventures.

Bobcats have been found to make interesting pets when raised from kittens, but are inclined to be very hard on the furniture. In the wild they are preyed upon chiefly by cougars, wolves, and coyotes, in addition to man.

OCELOT
LEOPARDUS PARDALIS

The prime source of danger for this New World felid is the demand for its beautiful fur. Once a rather common animal in Central and South American forests and scrublands, the ocelot is presently in danger of becoming extinct.

Ocelots feed on mice, rats, guinea pigs, monkeys, pacas, agoutis, rabbits, and small deer; sometimes, poultry is taken. Between 25 and 40 inches in length, excluding the tail, the ocelot is an excellent hunter and, like many cats, unafraid of water. It is also a good swimmer. Ocelots often live in pairs, hunting together and maintaining contact with each other by calling back and forth.

Although hand-raised ocelots can be very tame and dependent on their owners, they are not suitable house pets because they possess a strong body odor, and both males and females commonly spray.

The jaguar is the largest cat native to the western hemisphere. Ranging from central Patagonia in South America, it has, on occasion, been found as far north as the southwestern United States. In size and marking, it looks very much like a leopard, although the jaguar is a much heavier animal, weighing up to 300 pounds.

It can be distinguished from a leopard also by its dark markings, which are arranged in rosettes of four or five around a central light spot. The natural prey of this carnivore includes a large variety of animals of tropical South America. The more important of these are the capybara—the world's largest rodent—peccaries, deer, tapirs, turtles, and alligators.

Inhabiting thick wooded country and arid shrubby areas, jaguars often make their dens in caves or under overhanging rocks. Their relatively heavy bodies prevent these cats from being good climbers, but, on the other hand, they are good swimmers. The usual litter consists of one to four young, born about 100 days after mating.

JAGUAR
PANTHERA ONCA

COUGAR
FELIS CONCOLOR

Next to the bears, this is the largest and most powerful predator in the United States and Canada. It was once numerous from the Atlantic to the Pacific but the advance of civilization has reduced its range until today, in North America, it is found chiefly in the Rocky Mountains from Canada to Mexico.

Large males can measure eight feet in total length and can weigh over 200 pounds. They prey upon deer, wapiti, bighorn sheep, mountain goats, beaver, showshoe hares, and even mice. Their preference is for fresh meat and they rarely scavenge from old kills.

Due to their extreme wariness, these animals are not often seen by man in their natural habitat, despite their size.

BLACK BEAR
URSUS AMERICANUS

The black bear ranges all across forested North America, from central Mexico to Canada.

Solitary animals most of the year, they pair up briefly during the mating season. Cubs remain with their mother for about a year, and they frequently owe their lives to her fierce protective instinct, which prevents them from being killed by the adult males, who can be surly and bad-tempered at times.

Black bears swim well and often climb trees to feed on buds. They have a keen sense of smell and acute hearing, but poor eyesight. They can be seen at any hour of the day but are most active at night.

When very young, the cubs cry when afraid and hum when contented.

Black bears are omnivorous, their diet consisting of about 75 percent vegetable matter (including fruits and tender grasses), 15 percent carrion and the remainder being insects and small mammals. Their love for honey is well known, and in autumn, sweet, ripe corn also attracts them.

They have few natural enemies, but the one they fear the most is the grizzly bear. Whenever their territories overlap, the latter is given a wide berth.

GRIZZLY BEAR
URSUS ARCTOS

The grizzly bear has the reputation of being the most ferocious and dangerous mammal in North America. Some types are rangy and others more stockily built; some have long, slender heads, and others have short, broad heads. Similarly, there are wide color variations, the tundra grizzly often being creamy yellow on the back with brownish legs and underparts. In the Rocky Mountains the "silver-tip" phase is dominant. Adults weigh from 300 to 1,150 pounds, and are prodigiously strong.

Grizzlies will for the most part avoid contact with man, but they are sometimes unpredictable and should be given plenty of room. They move with a slow, shambling walk, the low-slung head swinging from side to side. But when the urge is on them, they can move very quickly and even horses are hard put to evade the bear's initial rush.

This powerful animal once inhabited almost all of western North America, but with the advent of Europeans its numbers were reduced. Now it is restricted chiefly to the northern Rockies and Alaska.

POLAR BEAR
URSUS MARITIMUS

One of the earth's largest and most powerful carnivores, the polar bear has a longer neck than other bears, and a long, narrow head. Its legs are also much longer. It inhabits all the world's arctic seas and coastlines.

When disturbed, it heads for open water. It moves in a rolling gallop, and can reach a top speed of about 25 miles per hour. The polar bear may surpass the Alaskan grizzly in size; the largest specimen recorded weighed 1,600 pounds. Adult males often weigh between 900 and 1,100 pounds. It is the most carnivorous of the bears, its favorite prey being seals and young walrus. In addition to its mainly meat diet, it grazes on grasses, mushrooms, and berries.

Aside from man, polar bears have no natural enemies, although they may occasionally fall victim to killer whales.

BIG BROWN BAT
EPTESICUS FUSCUS

Despite its name this is only a medium-sized bat, about five inches long with a wingspread of 12 or 13 inches. The fur is rather oily in texture and rusty brown in color. The wing membranes are almost black and of a leathery texture.

This species has become well adapted to civilization and is commonly observed at dusk flying along lamplit streets in pursuit of insects attracted to the light. During the summer months these bats often seek shelter behind loose boards, under eaves or shingles, in attics and church belfries, and so on.

Many spend the winter months in caves, sometimes hanging in tightly packed clusters of 100 or more individuals.

These creatures have few natural enemies but are a nuisance to man because of their habit of occupying buildings. In addition, they sometimes carry rabies.

PRONGHORN ANTELOPE
ANTILOCAPRA AMERICANA

Once roaming the prairies in numbers rivalling the bison, the pronghorn is now restricted, in Canada, to the adjacent southern corners of Alberta and Saskatchewan. In the United States, it is found in the grasslands and rocky deserts of the west from Montana to Arizona. A small, trim animal standing about three feet high at the shoulder, the pronghorn is the fastest North American mammal, capable of short bursts of speeds estimated at 50 to 60 miles per hour.

Pronghorns travel in small herds, and signal one another by raising the white hair on their rumps, which flashes in the sunlight for long distances on the prairies.

Their chief predator is the coyote, while newborn kids sometimes fall victim to the Golden Eagle.

MUSKOX
OVIBOS MOSCHATUS

This shaggy wild relative of sheep and goats lives in herds on the tundra. The muskox is native to northern Canada, Greenland, and Alaska's north coast.

The Canadian population is estimated at about 10,000 animals, most of which live on the High Arctic islands.

The thick undercoat of fine wool is covered by long guard hairs and protects the muskox during the long, cold winters.

Adult males have massive horn bases, which are used in head-on clashes during fights over herd leadership. The sharp horn tips of the adults are used primarily in defense against wolves—their only predatory enemy besides man.

MOUNTAIN GOAT
OREAMNOS AMERICANUS

The mountain goat is found chiefly in the northern Rockies and especially British Columbia, where it inhabits rugged, mountainous terrain from Colorado to the Yukon and Alaska. Despite the fact that billies are considerably larger than nannies (averaging about 180 pounds), the social order is matriarchal, with the nannies dominating the more placid billies except during the mating season.

These animals are very adept at moving about on high rocky cliffs and ledges, where they spend much of their time. Their main predator is the cougar, and the kids sometimes fall victim to the Golden Eagle.

They are hunted by man as a trophy animal, since the meat is not considered very palatable by human standards.

BIGHORN SHEEP
OVIS CANADENSIS

This stocky, muscular animal is found in western North America, in the Rocky Mountain states of the United States, and in western Alberta and southern and eastern British Columbia in Canada. Adult males range in weight from 287 to 343 pounds. The ewes, lambs, yearlings, and young rams band together in groups led by an old ewe, and stay together all year. They are joined in autumn by the mature rams in time for the mating season, but the old ewe's leadership is not challenged by the males.

During the rutting season the adult males engage in fierce jousting matches, crashing head-on into each other with an impact that can be heard a mile away.

Chief predators are the cougar and the Golden Eagle, which is always alert for unattended kids.

Bighorn heads, with their magnificent horns, are prized trophies of the hunting fraternity.

ELK
CERVUS ELAPHUS

The American elk was named by early English settlers, but some people prefer to call it wapiti, its Shawnee Indian name, meaning "white rump."

This impressive-looking member of the deer family is second in size only to the moose. The average weight of an adult male is 700 pounds. Some individuals attain weights of 1,100 pounds.

In summer the wapiti's coat is sleek and tawny brown, with a large buff-colored rump patch surrounding the tail and including the buttocks. Elk are the most vocal members of the deer family. The bugling of the stags is a form of challenge and carries for about a mile on a clear day. Good swimmers, they can also attain speeds up to 30 miles per hour in a swift gallop. Once found in most of North America as far east as southeastern Quebec and the Allegheny Mountains, it has now been restricted to isolated pockets of unexploited habitat.

PEARY CARIBOU
RANGIFER TARANDUS PEARYI

The Peary caribou, smaller and lighter than the barren-ground caribou, lives in small herds on the Arctic islands.

Though Peary caribou regularly move between summer and winter ranges, and may travel between islands over the winter ice, they do not make the spectacular migrations for which the barren-ground caribou are well known.

A thick white coat of hollow hairs provides good insulation from winter temperatures.

To obtain food in winter, caribou dig feeding craters in the snow, pawing down to the vegetation below with their broad hooves. The antlers of the Peary caribou grow swiftly each year, and may double the height of an adult male before growth ends in the fall. After the fuzzy velvet is stripped off (sometimes in sparring matches which precede the rut), the antlers of both males and females become a pure, shining white.

MOOSE
ALCES ALCES

The largest member of the deer family, this animal is found in northern North America from Alaska to Newfoundland and Maine, and in northern Europe from Scandinavia eastward to the Pacific coast. An adult bull averages about 1,000 pounds in weight, with a larger subspecies from Alaska and the Yukon attaining weights in excess of 1,500 pounds.

The moose is a strong swimmer and has been known to dive to depths of 18 feet and remain submerged for 30 seconds in search of underwater vegetation.

An adult bull is a dangerous adversary, often able to repel attacks by small packs of wolves. They are particularly truculent during the mating season.

Possessed of keen senses of smell and hearing, they can be approached only from upwind.

BISON
BISON BISON

No one knows how many bison inhabited North America before the coming of the white man. Most estimates range from 50 to 100 million animals. They were the economic cornerstone of the plains Indians, who utilized almost every part of the animal to provide food, shelter, weapons, utensils, and ornamentation.

A large bull may stand six feet high at the shoulder and weigh 2,000 pounds. Among the predators, only the grizzly bear is strong enough to kill an adult bull. Bison are equipped with a keen sense of smell, capable of detecting odors up to a mile away. They also have excellent eyesight and are quite at ease in the water. They have been known to live for up to 40 years.

WALRUS
ODOBENUS ROSMARUS

This ponderous animal was once found as far south as the Gulf of St. Lawrence and the Massachusetts coast. Today it inhabits the edge of the Arctic ice sheet, appearing only as far south as James Bay and the Labrador coast. The average weight of an adult male is about 1,675 pounds but some of the large Pacific bulls can attain weights of almost 2,975 pounds.

Awkward and slow-moving on land, they are quite graceful and more belligerent in the water. Wounded bulls have been known to attack hunters' boats.

The walrus has long been a major resource of the Eskimo people, who use their hides for boat coverings and for thongs and dog traces. The bones and tusks are used for weapons, tools, and carvings.

The animal's vocabulary consists of grunts and bellows; the latter sometimes carry for a mile or more.

Aside from man, their principal enemies are the polar bear and the killer whale.

NORTHERN ELEPHANT SEAL
MIROUNGA ANGUSTIROSTRIS

This occasional visitor to the northern Pacific waters from Canada to Alaska breeds in California and is one of the world's largest living mammals. Adult bulls may grow to 19 feet in length and a weight of 7,716 pounds has been recorded. Aside from its dimensions, the most striking aspect of this ponderous, lethargic creature is the foot-long proboscis sported by adult males. When inflated, it curves downward into the mouth. Bellows can carry up to a mile and are a prime source of intimidation of its adversaries in the herd.

Elephant seals are diurnal in activity, but spend a good deal of the daylight hours sleeping on the beaches. They are strangely indifferent to man and are easily approached.

Their diet consists of squid, fish, and small sharks. They, in turn, are sometimes preyed upon by killer whales.

HOODED SEAL
CYSTOPHORA CRISTATA

The hooded seal is so named because of its large, elastic nasal cavity, or "hood," which extends from its nostrils to its forehead. When fully inflated, it resembles a large black rubber ball. It is lacking in females and immature males.

Uncommonly aggressive for seals, these are large animals, males occasionally exceeding 10 feet in length and 880 pounds in weight. They have been reported to live for as long as 32 years.

Most of the hooded seal population is distributed in the north Atlantic, including the waters around Canada's Maritime provinces, Newfoundland, and into the Gulf of St. Lawrence. Their diet consists of mussels, starfish, squid, shrimp, herring, and cod. Aside from man, who hunts them for their skin, their chief natural predator appears to be the killer whale.

HARP SEAL
PHOCA GROENLANDICA

This is probably the most important of the seals from the commercial point of view.

Adults average about six feet in length; maximum weight runs up to 400 pounds.

The pups, at birth, are covered with a long fluffy white fur, from which they derive their common name of "whitecoats." Mothers find their own pups among the many whitecoats on the ice floes by odor, and reject all but their own.

The pups nurse for only about two weeks and are then abandoned to fend for themselves. During that two-week period, however, they grow enormously on their mother's milk, which is rich in butterfat. When weaned, pups weigh between 90 and 100 pounds.

Harp seals feed primarily on small fish and crustaceans. They are reported to be capable of diving to depths of up to 150 fathoms, and of remaining submerged for as long as 15 minutes. They apparently live for 30 years or more. Aside from humans, their principal predators are Greenland sharks, killer whales, and polar bears.

GRAY WHALE
ESCHRICHTIUS ROBUSTUS

This large whale, which can grow to lengths of up to 50 feet, was nearly exterminated by the Pacific whaling industry during the late 1800s. It came to an early extinction in the north Atlantic and may be close to extinction on the Asian side of the Pacific. Since their protection by international agreement in 1937, they have increased on the North American side, where they range from their summer waters in the Bering and Chukchi Seas to winter waters off Baja California. The estimated population is 11,000 animals, which, it is hoped, may re-stock the Japanese and Korean waters.

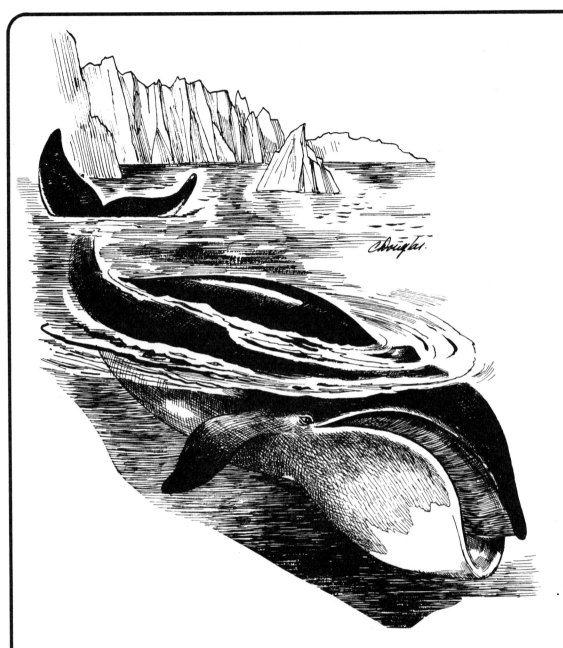

BOWHEAD WHALE
BALAENA MYSTICETUS

This slow-moving whale, which can grow to lengths of up to 65 feet, inhabits the Arctic Ocean and subarctic waters north of the Atlantic and Pacific oceans.

Whaling for this species began as early as 1611, near Spitsbergen, and continued until the early 1900s when the numbers became too low for economic interest and when the demand for whalebone ceased. The bowhead, close to extinction, has been protected from modern whaling since 1937 by both its low numbers and the International Whaling Commission. Some may be taken for food by native peoples. Recovery seems to be going well in the Bering, Chukchi, and Beaufort Seas, but is apparently slower in the Baffin Island, Greenland, and Spitsbergen regions, where whaling was more intensive and was carried on over a longer period.

FIN WHALE
BALAENOPTERA PHYSALUS

This whale is sometimes called "the greyhound of the deep." The average length of North Pacific males is 62 feet while those found in the southern hemisphere may approach 82 feet. They can reach speeds of up to 25 knots for short intervals, and usually appear in small groups of two to seven individuals. An adult fin whale may consume 2,200 to 3,300 pounds of food per day, feeding mainly on krill, but occasionally consuming herring as well.

In the Atlantic, fin whales reach the Gulf of St. Lawrence on their northern migration in March, and are common off Newfoundland in June, usually appearing about 25 miles offshore. They are also fairly common off the West Coast most of the summer. Present numbers appear to be declining.

HUMPBACK WHALE
MEGAPTERA NOVAEANGLIAE

Distinguished by its short, stout body and long, curved flippers, which are often as long as a third of the total body length, this slow-moving animal is usually docile and easily approached. Average lengths are 41 feet for Pacific adult males and 48 feet for females. The average weight is 30 tons.

The humpback is often observed throwing itself out of the water in gigantic somersaults. It rolls on the surface, or sometimes leaps out of the water, flippers beating the air as though it were attempting to fly. In the mating season they sometimes use their flippers to administer love pats to each other, some of which are audible for miles.

Humpbacks occur in both hemispheres, migrating in summer to feed in the polar seas and retreating in winter to tropical waters to breed.

In addition to man, humpbacks are preyed upon by killer whales.

SEI WHALE
BALAENOPTERA BOREALIS

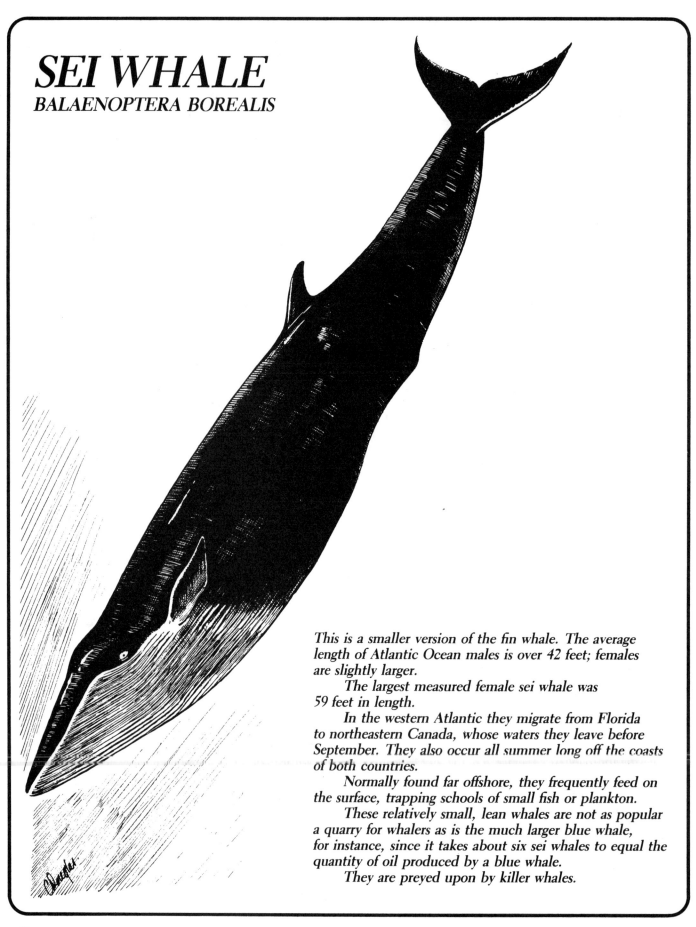

This is a smaller version of the fin whale. The average length of Atlantic Ocean males is over 42 feet; females are slightly larger.

The largest measured female sei whale was 59 feet in length.

In the western Atlantic they migrate from Florida to northeastern Canada, whose waters they leave before September. They also occur all summer long off the coasts of both countries.

Normally found far offshore, they frequently feed on the surface, trapping schools of small fish or plankton.

These relatively small, lean whales are not as popular a quarry for whalers as is the much larger blue whale, for instance, since it takes about six sei whales to equal the quantity of oil produced by a blue whale.

They are preyed upon by killer whales.

BLUE WHALE
BALAENOPTERA MUSCULUS

The largest creature ever known to have existed on this planet is the present-day blue whale. The largest dinosaur attained a length of about 72 feet and weighed about 36 tons. Today's blue whale may reach a total length of 100 feet and weigh 145 tons.

This giant animal is placid and shy by nature. On the surface, its normal cruising speed is about 12 knots, but it is capable of attaining 20 knots for short intervals. The maximum reported depth reached by the species is 194 fathoms, and it is capable of remaining submerged for 50 minutes, but 10 to 15 minutes is the usual time beneath the surface. The life span of a blue whale is about 30 years.

The calves measure 23 feet at birth and weigh about 4,400 pounds. By the time they are a year old the youngsters measure about 58 feet.

Blue whales comprised about 90 percent of the whaling industry's total catch during the early part of this century. In 1931 more than 30,000 of these majestic creatures were killed. Since then, the blue whale population has declined. By the year 1971, despite belated conservation measures, only 3,000 blue whales were thought to exist in the world's seas.

SPERM WHALE
PHYSETER CATODON

size comparison—man and sperm whale

The sperm whale, or cachalot, is one of the cetaceans, marine mammals whose ancestors were probably land animals.

Large males attain a length of 60 feet and can weigh as much as 50 tons. They range across most of the world's oceans, but are most usually found in those latitudes inhabited by giant squid, their favorite prey.

The whale sometimes descends to depths of 3,000 feet in search of these creatures, who themselves often measure over 30 feet from tip to tip at their tentacles.

In addition to squid, these huge animals also take seals, rays, and sometimes sharks up to 10 or 12 feet long.

Sperm whales have no teeth in their upper jaw, but those in the lower jaw, about 60 in number, are about eight inches long and weigh up to six or seven pounds each.

Found mainly in Arctic waters, the beluga, or white whale, journeys into the Gulf of St. Lawrence and other warmer waters to give birth.

In 1906 the skull and part of the skeleton of a young white whale were found in a well excavation near Pakenham, Ontario. They were buried beneath 14 feet of clay laid down in the Champlain Sea, an arm of the Atlantic Ocean which covered this area about 11,000 years ago.

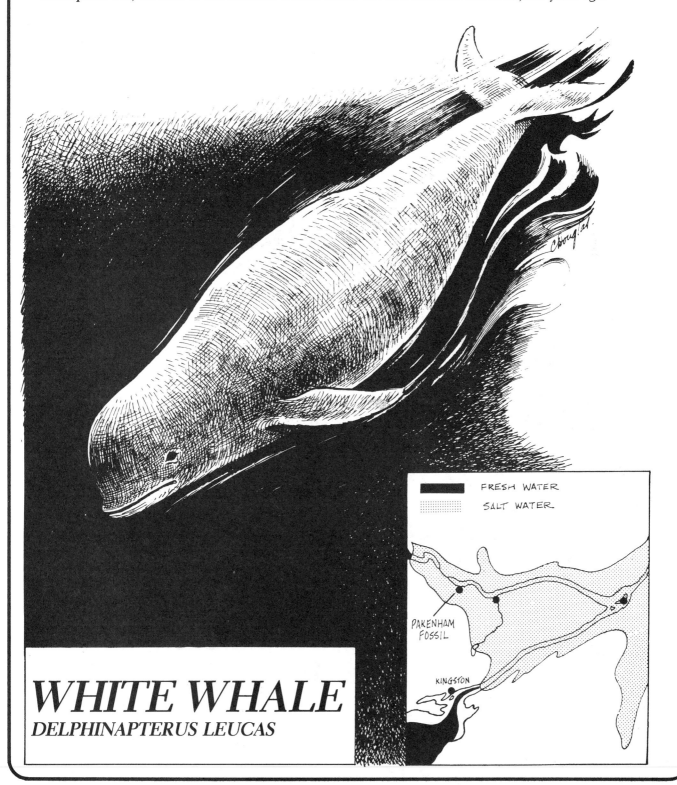

WHITE WHALE
DELPHINAPTERUS LEUCAS

Like other whales, the orca is a warm-blooded, air-breathing mammal. They are called "killers" because they prey on fish, dolphins, seals, and even larger whales. There are no documented cases of deliberate attacks on humans. The story of an orca found with 13 porpoises and 14 seals in its stomach has been officially discredited. Orcas travel and hunt in social groups (pods) of three to 40 whales, throughout most of the world's oceans.

Adult males are recognized by their tall, triangular dorsal fins. Females and immature males have a smaller, curved fin.

Large bulls can attain lengths of up to 30 feet and can weigh up to eight tons.

ORCA, or KILLER WHALE
ORCINUS ORCA

FLORIDA MANATEE
TRICHECHUS MANATUS LATIROSTRIS

Although recorded in the Carolinas, Georgia and Texas, the manatee's center of abundance in the United States was, and is, Florida, where it frequents streams and brackish estuaries and feeds solely on vegetation. Early depletion of this aquatic mammal began with hunting by Indians for hides, meat, and oil. Later, the loss of habitat to coastal development, the use of herbicides to control aquatic vegetation, and injuries from power boats have all contributed to its destruction.

Although now legally protected, the animal's numbers are low as it continues to survive in Everglades National Park and a refuge near Tampa. Other populations of manatees in the coastal Caribbean are probably declining as well.

BIBLIOGRAPHY: SELECTED READING LIST

HERPETOLOGY (AMPHIBIANS AND REPTILES)

Bellairs, A., and R. Carrington
(1966). *The World of Reptiles.* London, England: Chatto and Windus.

Cochran, D. M.
(1961). *Living Amphibians of the World.* Garden City, N.Y.: Doubleday and Co.

Conant, R.
(1975). *A Field Guide to Reptiles and Amphibians of Eastern and Central North America.* 2nd ed. Boston: Houghton Mifflin.

Cook, F. R.
(1984). *Introduction to Canadian Amphibians and Reptiles.* Ottawa, Canada: National Museum of Natural Sciences, National Museums of Canada. (Distributed in the United States by University of Chicago Press.)

Goin, C. G., O. B. Goin, and G. R. Zug
(1978). *Introduction to Herpetology.* San Francisco: W. H. Freeman and Co.

Noble, G. K.
(1931). *Biology of the Amphibia.* New York: McGraw Hill. (Reprinted by Dover Books, 1954.)

Schmidt, K. P., and R. F. Inger
(1957). *Living Reptiles of the World.* Garden City, N.Y.: Doubleday and Co.

Smith, H. M., and E. D. Brodie, Jr.
(1982a). *Amphibians of North America: A Guide to Field Identification.* New York: Golden Press.
(1982b). *Reptiles of North America: A Guide to Field Identification.* New York: Golden Press.

Stebbins, R. C.
(1966). *A Field Guide to Western Amphibians and Reptiles.* Boston: Houghton Mifflin.

ICHTHYOLOGY (FISHES)

Grzimek, B.
(1973–1974). *Grzimek's Animal Life Encyclopedia.* Volumes 4 and 5. New York: Van Nostrand Reinhold.

Hart, J. L.
(1973). *Pacific Fishes of Canada.* Ottawa, Canada: Fisheries Research Board of Canada, Bulletin 180.

Lee, D. S., et al.
(1980). *Atlas of North American Freshwater Fishes.* Raleigh: North Carolina State Museum of Natural History.

Leim, A. H., and W. B. Scott
(1966). *Fishes of the Atlantic Coast of Canada.* Ottawa, Canada: Fisheries Research Board of Canada, Bulletin 155.

McAllister, D. E., and E. H. Crossman
(1973). *A Guide to the Freshwater Sport Fishes of Canada.* Ottawa, Canada: National Museum of Natural Sciences, National Museums of Canada.

Ono, R. D., J. D. Williams, and A. Wagner
(1983). *Vanishing Fishes of North America.* Washington, D.C.: Stone Wall Press.

Scott, W. B., and E. J. Crossman
(1973). *Freshwater Fishes of Canada.* Ottawa, Canada: Fisheries Research Board of Canada, Bulletin 184.

Sterba, G.
(1983). *The Aquarium Encyclopedia.* Cambridge, Mass.: The MIT Press.

MAMMALOGY (MAMMALS)

Banfield, A. W. F.
(1974). *The Mammals of Canada.* Toronto, Canada: University of Toronto Press.

Burt, W. H., and R. P. Grossenheider
(1952). *A Field Guide to the Mammals.* Peterson Field Guide Series. Cambridge, Mass.: Riverside Press.

Grzimek, B.
(1972–1975). *Grzimek's Animal Life Encyclopedia.* Mammals: volumes 10, 11, 12 and 13. New York: Van Nostrand Reinhold.

Hamilton, W. H., Jr., and J. O. Whitaker, Jr.
(1979). *Mammals of the Eastern United States.* 2nd ed. Ithaca, N.Y.: Cornell University Press.

Peterson, R. L.
(1966). *The Mammals of Eastern Canada.* Toronto, Canada: Oxford University Press.

van Zyll de Jong, C. G.
(1983). *Handbook of Canadian Mammals.* Volume 1: Marsupials and Insectivores. Ottawa, Canada: National Museum of Natural Sciences, National Museums of Canada.

(1985). *Handbook of Canadian Mammals.* Volume 2: Bats. Ottawa, Canada: National Museum of Natural Sciences, National Museums of Canada.

Walker, E. P.
(1975). *Mammals of the World.* 3rd ed. Volumes 1 and 2. Baltimore: Johns Hopkins University Press.

Whitaker, J. O., Jr.
(1980). *The Audubon Society Field Guide to North American Mammals.* Toronto, Canada: Random House.

ORNITHOLOGY (BIRDS)

Allen, T. B.
(1974). *Vanishing Wildlife of North America.* Washington, D.C.: National Geographic Society.
(1976). *Our Continent: A Natural History of North America.* Washington, D.C.: National Geographic Society.

Bull, J. L., and J. Farrand
(1977). *The Audubon Society Field Guide to North American Birds, Eastern Region.* New York: Knopf.

Dorst, J.
(1974). *The Life of Birds.* New York: Columbia University Press.

Peterson, R. T.
(1972). *Field Guide to Western Birds.* Boston: Houghton Mifflin.
(1980). *A Field Guide to the Birds East of the Rockies.* 4th ed. Boston: Houghton Mifflin.

Time-Life Books
(1977). *Birds of Field and Forest.* New ed. New York: Time-Life Inc.
(1978). *Songbirds.* New York: Time-Life Inc.

Udvardy, M. D. F.
(1977). *The Audubon Society Field Guide to North American Birds, Western Region.* New York: Knopf.

PALEOBIOLOGY (DINOSAURS)

Case, G. R.
(1982). *A Pictorial Guide to Fossils.* New York: Van Nostrand Reinhold.

Charig, A., and B. Horsfield
(1975). *Before the Ark.* London, England: British Broadcasting Corporation.

Colbert, E. H.
(1961). *Dinosaurs: Their Discovery and Their World.* New York: E. P. Dutton.
(1968). *Man and Dinosaurs.* New York: E. P. Dutton.
(1983). *Dinosaurs: An Illustrated History.* Maplewood, N.J.: Hammond Inc.

Fortey, R.
(1982). *Fossils: The Key to the Past.* London, England: Heinemann.

Glut, D. F.
(1982). *The New Dinosaur Dictionary.* Secaucus, N.J.: Citadel Press.

Kurtén, B.
(1971). *The Age of Mammals.* London, England: Weidenfeld and Nicolson.

Kurtén, B., and E. Anderson
(1980). *Pleistocene Mammals of North America.* New York: Columbia University Press.

Romer, A. S.
(1966). *Vertebrate Paleontology.* 3rd. ed. Chicago: University of Chicago Press.

Russell, D. A.
(1977). *A Vanished World: The Dinosaurs of Western Canada.* Ottawa, Canada: National Museum of Natural Sciences, National Museums of Canada.

INDEX

A

Alligator
　American, 62
Antelope
　Pronghorn, 136
　Saiga, 17
Ass
　Yukon Wild, 16
Auk
　Great, 32

B

Badger
　American, 119
Bat
　Big Brown, 135
Bear
　Black, 132
　Grizzly, 133
　Polar, 134
　Short-faced, 26
Beaver, 105
　Giant, 33
Bison, 143
Bobcat, 128
Brachiosaurus, 12
Bullfrog, 51
Bullhead
　Brown, 44
Bunting
　Snow, 86

C

Camel
　Western, 15
Caribou
　Peary, 141
Carp, 41
Cat
　Scimitar, 24
Chipmunk
　Eastern, 96
Condor, 76
Cougar, 131
Coyote, 124
Crane
　Whooping, 74
Crocodile, 61
Crow
　American, 89
Curlew
　Eskimo, 91

D

Dawn Horse, 14

Dogs
　earliest in North
　　America, 29
Dovekie, 68
Dromiceiomimus, 13

E

Eagle
　Bald, 81
　Golden, 80
Elk, 140
Eohippus, 14
Ermine, 116
Euoplocephalus, 8

F

Falcon
　Peregrine, 79
Ferret
　Black-footed, 117
Fish
　Lobe-finned, 3
Fox
　Red, 123
Frog
　Northern Leopard, 49
　Striped Chorus, 50

G

Gannet
　Northern, 67
Goat
　Mountain, 138
Goose
　Canada, 72

H

Hare
　Arctic, 108
Herom
　Great Blue, 75
Horse
　Dawn, 14
Hummingbird
　Ruby-throated, 88
Hypacrosaurus, 5

I

Ice-age animals on
　Vancouver Island, 19

J

Jaguar, 130

L

Lamprey
　Sea, 43
Lion
　American, 28
Lizard
　Alligator, 63
Lynx, 127

N

Mammoth
　Babine Lake, 23
　Woolly, 25
Manatee
　Florida, 157
Marmot
　Hoary, 107
　Vancouver Island, 103
Marten
　American, 118
Mastodon
　American, 22
Mink
　American, 114
Mole
　Hairy-tailed, 111
　Star-nosed, 110
Moose, 142
Mouse
　House, 100
Murre
　Common, 66
Muskox, 137
　Helmeted, 18
Muskrat, 104

N

Nighthawk
　Common, 84

O

Ocelot, 129
Orca, 156
Otter, 112
　Sea, 113
Owl
　Burrowing, 77
　Snowy, 78

P

Panoplosaurus, 11
Parakeet
　Carolina, 31

Pelican
　American White, 71
Pigeon
　Passenger, 30
Pike
　Northern, 42
Plover
　Piping, 82
Porcupine
　American, 109
Prairie-Chicken
　Greater, 70
Puffin
　Atlantic, 69

R

Racer, 52
Raccoon, 122
Rat
　Brown, 102
　Kangaroo, 101
Rattlesnake
　Massasauga, 53
Raven
　Common, 90
Razorbill, 66
Reptiles
　Coal-Age, 4
　Flying, 10
Ridley
　Atlantic, 58
Robin
　American, 92

S

Salamander
　Yellow-spotted, 48
Salmon
　Atlantic, 46
Sapsucker
　Yellow-bellied, 83
Seal
　Harp, 147
　Hooded, 146
　Northern Elephant, 145
Seals
　Ringed, 21
Shark
　Blue, 40
　Blue Pointer, 38
　Great White, 38
　Greenland, 38
　Pacific Mako, 38
　Smooth Hammerhead, 38
　Thresher, 38

Whale, 39
Sharks, 38
Sheep
　Bighorn, 139
Shrike
　Northern, 85
Skunk
　Striped, 121
Sloth
　Jefferson's Ground, 27
Snake
　Black Rat, 55
　Common Garter, 54
Squirrel
　Northern Flying, 97
　Red, 99
　Richardson's Ground, 98
Stenonychosaurus, 6
Sturgeon
　Shortnose, 45
Styracosaurus, 7
Swallow
　Barn, 87
Swan
　Trumpeter, 73

T

Trout
　Brown, 47
Turtle
　Atlantic Loggerhead, 59
　Common Snapping, 57
　Leatherback, 60
　Spiny Softshell, 56
Tyrannosaurus, 9

W

Walrus, 144
Warbler
　Kirtland's, 93
Weasel
　Least, 115
Whale
　Blue, 153
　Bowhead, 149
　Fin, 150
　Gray, 148
　Humpback, 151
　Killer, 156
　Sei, 152
　Sperm, 154
　White, 155
　White Lake, 20
Wolf, 125
　Arctic, 126
Wolverine, 120